Developing Researcher Independence Through the Hidden Curriculum

Dely L. Elliot
Søren S. E. Bengtsen • Kay Guccione
Editors

Developing Researcher Independence Through the Hidden Curriculum

palgrave
macmillan

Editors
Dely L. Elliot
University of Glasgow
Glasgow, UK

Søren S. E. Bengtsen
Aarhus University
Aarhus, Denmark

Kay Guccione
University of Glasgow
Glasgow, UK

ISBN 978-3-031-42874-6 ISBN 978-3-031-42875-3 (eBook)
https://doi.org/10.1007/978-3-031-42875-3

This Palgrave Macmillan imprint is published by the registered company Springer Nature Switzerland AG.
The registered company address is: Gewerbestrasse 11, 6330 Cham, Switzerland

Paper in this product is recyclable.

We owe the various ideas and concepts, research evidence, lessons and stories presented in this book to many individuals we worked and interacted with over the years. You have inspired, taught and encouraged us to continue reflecting on doctoral pedagogies so that together, we can foster a more enriching and transformative doctoral learning experience not only for our doctoral scholars, but also for everyone who supports them in their journey—both inside and outside academia.

We believe that behind every successful doctorate, there stands a strong interdependent and supportive community!

Contents

Notes on Contributors

Melina Aarnikoivu is a postdoctoral researcher at the Finnish Institute for Educational Research, University of Jyväskylä, Finland. Her research interests include doctoral education, academic writing, and community-building. Melina is also one of the coordinators of the Early Career Higher Education Researchers (ECHER) network.

Claire Aitchison has worked, researched, and published in higher education since 1991. She is an academic developer in the Teaching Innovation Unit at the University of South Australia. Claire has a passion for enabling pedagogies that build skills and knowledge through writing as a social practice. She has published widely on doctoral writing, including edited/co-authored books, book chapters and journal papers. She initiated the scholarly blog DoctoralWriting which she co-edits with Cally Guerin and Susan Carter.

Ruth M. Albertyn is an extraordinary associate professor affiliated with the Centre for Higher and Adult Education and the Business School at Stellenbosch University. She has been involved in the higher education sector for three decades, has supervised and examined master's and doctoral theses, and mentors early career academics. She presents capacity development workshops for academics at universities in Africa and Belgium. Ruth has won awards for teaching, research, review, and a National Research rating recognising her expertise in postgraduate research development. She has co-edited three scholarly books in the higher education field.

Dayana Balgabekova is a doctoral researcher in the School of Education at the University of Glasgow. Her PhD thesis explored the professional learning of teachers in the context of Lesson Study in Kazakhstan. Dayana's research interests include professional development, mentoring, development of twenty-first-century skills and innovative research methods.

Sarah Barradell is a senior lecturer in the Department of Nursing and Allied Health at Swinburne University, Victoria, Australia, and is a physiotherapist with extensive clinical experience in neurological physiotherapy. Her research interests are centred around health professional education, helping students learn for changing practice and with more influence, staff-student partnerships, and participation in the scholarship of teaching and learning. Sarah is a nationally awarded teacher, and a fellow, Advance Higher Education, and is serving as an associate editor for *Physiotherapy Theory and Practice* and *Higher Education Research and Development.*

Amani Bell joined the School of Health Sciences at The University of Sydney, Australia in 2019 after her role as Vice-Chancellors' Fellow at the Innovative Research Universities, where she led a suite of projects exploring models of work-integrated learning that are responsive to the changing nature of work. With a long background in academic development, Amani's research focuses on students and staff working in partnership to enhance higher education, particularly to improve access and success for students from diverse backgrounds. Her co-edited book *Understanding Experiences of First Generation University Students: Culturally Responsive and Sustaining Methodologies* was published in 2018 (Bloomsbury).

Søren S. E. Bengtsen is an associate professor at the Department of Educational Philosophy and General Education, Danish School of Education, Aarhus University, Denmark. At Aarhus University, he is the co-director of the research centre, 'Centre for Higher Education Futures' (CHEF). He serves as Chair of the Philosophy and Theory of Higher Education Society (PaTHES) and as vice-chair of the Danish Network for Educational Development in Higher Education (DUN). He is the co-editor of the two book series: 'Debating Higher Education: Philosophical Perspectives' (Springer) and 'World Issues in the Philosophy and Theory of Higher Education' (Routledge).

Maria Berge is Associate Professor of Science and Technology Education at Umeå University, Sweden. Her research interests are centred around issues related to interactional patterns and norms in the context of learning science and technology. She is part of three research projects: one that concerns how the figuration of 'the geek' includes and excludes, one about how engineering students express and deal with emotions when they discuss sustainability problems, and one where processes of inclusion and exclusion in higher education mathematics and physics are explored.

Billy Bryan is an evaluation and research leader in the Science and Emerging Technologies (SET) team at RAND Europe, a not-for-profit policy thinktank based in Cambridge, UK. He has conducted research and written on the value of the doctorate in the modern knowledge economy for the past 8 years with his Co-I, Dr Kay Guccione. He is also a trustee at the University of Sussex Students' Union, an advisor to the Cardiovascular Research Trust (CVRT) and a member of the UK Government's Evaluation and Trial Advice Panel (ETAP).

Kun Dai is an assistant professor at the Department of Educational Administration and Policy, Faculty of Education, The Chinese University of Hong Kong. His research interests include international and comparative education, transnational higher education and education policy. Dr Dai's work has appeared in several leading international peer-reviewed journals, such as *Higher Education, Studies in Higher Education, Higher Education Research & Development, Assessment & Evaluation in Higher Education* and *Teaching in Higher Education.* Dr Dai serves as an associate editor of the *Journal of International Students,* an editorial board member of *Compare* and the *Journal of Diversity in Higher Education.*

Dangeni is a professional development advisor at Newcastle University, where her teaching and research focus broadly on teaching and learning provision in the wider context of the internationalisation of higher education. She is particularly interested in research and practices around international students' access, engagement and success in postgraduate teaching (PGT) and postgraduate research (PGR) settings.

Dely Lazarte Elliot is a senior lecturer in the School of Education and a Quality Enhancement and Assurance Officer for the College of Social Sciences, University of Glasgow. Dely leads the Community Building team for the School of Education's doctoral community and co-leads UofG's Supervisor Community of Practice. She serves as the Deputy Editor for Innovations in Education and Teaching International and is on the Editorial Board for Higher Education Research and Development and Teaching in Higher Education. Dely is the author of a Routledge book entitled *Navigating Your International Doctoral Experience (and Beyond)* and the lead author of this book's predecessor *The Hidden Curriculum in Doctoral Education* (Palgrave).

Liezel Frick is a professor at the Centre for Higher Education within the Department of Curriculum Studies, and vice dean for Research and Postgraduate Studies at the Faculty of Education, Stellenbosch University, South Africa. She is also a research fellow at the Stellenbosch University DSI/NRF Centre of Excellence in Science, Technology and Innovation Policy (SCIStip). Her research interests include doctoral education and early career researcher development.

Karen Gordon is a doctoral researcher in The School of Education at the University of Glasgow. Her research focuses on distinct types of stress responses that doctoral researchers experience whilst completing their PhD. Her research interests lie in somatic and psychologised stress responses, cultural differences in mental health and creative research methodologies.

Owen Gower is the Director of the UK Council for Graduate Education, where he leads the Council's advocacy for postgraduate education and research. He has served on a number of national bodies, including: the UKRI Talent and Skills Advisory Group; Student Space which supports student mental health; Generation Delta—a project to increase the number of Black female professors in the UK; and the Equity in Doctoral Education Through Innovation and Partnership project. He is the author of the *UK Research Supervision Survey Report* (2021). He is a member of the Philosophy of Education Society of Great Britain, a Fellow of Goodenough College, and of the Royal Society of Arts.

Barbara M. Grant is an associate professor in the School of Critical Studies in Education at Waipapa Taumata Rau/University of Auckland. She conducts research in the field of critical university studies, with a particular interest in doctoral education, including the supervision of graduate students, academic work and identities, and activism within the university. She has also written about research methodologies and academic writing. She is writing a book about women doctoral supervisors in Aotearoa, New Zealand.

Kay Guccione is the Head of Research Culture & Researcher Development at the University of Glasgow. She is a Principal Fellow of the

Higher Education Academy and a National Teaching Fellow. Her recent projects and collaborations are the *Hidden Curriculum in Doctoral Education* (Palgrave Macmillan), *Coaching and Mentoring for Academic Development* (Emerald), *Thriving in Part-Time Doctoral Study* (Routledge), the 'Supervising PhDs' blog, and her work with the UK Council for Graduate Education to develop a Recognised Associate Supervisor Award.

Cally Guerin is a researcher developer in the Researcher Development Unit at The Australian National University. She has been teaching, researching, and publishing on doctoral education since 2008, working with both supervisors and HDRs to develop effective research practices. Much of her time is spent translating the mysteries of academic research cultures for newcomers. Her research interests include research writing, supervision practices and doctoral education. Along with Claire Aitchison, she co-edited *Writing Groups for Doctoral Education and Beyond: Innovations in Practice and Theory* (2014). She is Joint Executive Editor of *HERD* journal and a founding co-editor of Doctoral Writing.

Sally Hancock is Senior Lecturer in Education at the University of York, specialising in higher education research. Her research has focused on inequalities in access to postgraduate study; doctoral student aspirations and career intentions; employment outcomes; and the educational and economic policies shaping these issues. Recent projects include the analysis of doctoral employment outcomes in the UK, and a review of doctoral training in the social sciences. Her research has featured in *Nature*, the *Times Higher Education*, and informed guidance on doctoral supervision at a number of universities and through the UK Council for Graduate Education.

Jiawen Han holds a PhD from Ochanomizu University in Japan (Major: Applied Linguistics). Her research primarily centres around literacy education, translanguaging, and the development of literacy skills within the domains of reading and writing in the context of Japanese language acquisition and utilisation. She worked as a tutor for students who encounter problems with writing a research/report/paper at Ochanomizu University for several years. As a part-time lecturer, she taught Japanese writing at Ochanomizu University and a Japanese-language school in Tokyo. Recently, she has returned to China to continue her project concerning the literacy development of Chinese university students at Tianjin University of Technology.

Rui He is a lecturer at the Manchester Institute of Education (MIE) at the University of Manchester, UK. At MIE, she also serves as the Co-ordinator for Postgraduate Academic Enlightenment Programme. She is also the Founding Convenor of the Eastern Higher Education Research Network (ETHER) and the ChinaHE PGR/ECR Network and is part of the China and Higher Education Network Committee. Her research interests focus broadly on international education and cross-cultural psychology, particularly on international students' acculturation experiences and (im)mobility; mental health and psychological well-being; study abroad programmes; language, culture, and identity; and creative research methods.

Kelsey Inouye is a research associate at the Centre on Skills, Knowledge, and Organisational Performance, Department of Education, University of Oxford. She is also a supernumerary fellow at Jesus College, Oxford. Her research areas include doctoral education, researcher careers, and academic writing.

Giedre Kligyte is a senior lecturer in the TD School (Transdisciplinary School) at the University of Technology Sydney (UTS), Australia. Giedre's expertise is in transdisciplinary education where she has led several collaborative research projects examining how future-oriented capabilities can be fostered through transdisciplinary partnerships across organisational and disciplinary divides. She is the Program Director for the new Transdisciplinary Electives Program and teaches in the award-winning Bachelor of Creative Intelligence and Innovation and Masters of Creative Intelligence and Strategic Innovation degree programs.

Sofie Kobayashi is a lecturer at the University of Copenhagen and an independent consultant. Her research interests cover doctoral education and supervision, with a focus on how doctoral students develop independence, learn to do research, and navigate the norms of science. Her second strand of interest covers research and teaching in the development of teaching competences in higher education.

Delyse Leadbeatter is an associate professor and the Director of Academic Education at Sydney Dental School. Her interests in teaching and research are grounded in the idea that higher education is a complex social phenomenon. Delyse has a particular interest in researching inquiry learning and students as partners in all aspects of university endeavour. Her focus is curriculum co-design as a methodology for involving partners such as students, professionals and community representatives in the education of contemporary practitioners.

Kara A. Makara is a senior lecturer in the School of Education at the University of Glasgow, Joint Director of the University of Glasgow Educational Assessment Network and a member of the Robert Owen Centre for Educational Change. Her research interests include motivation and learning, social interactions within educational settings, academic help-seeking, assessment for formative purposes, and cultural and international differences in these processes. She is a Principal Investigator on the Camau i'r Dyfodol (Steps to the Future) Project, which brings together policymakers, teachers, researchers, and educational partners to co-construct knowledge to support curriculum realisation in Wales.

Catherine Manathunga is Professor of Education Research and Associate Dean (Research) in the School of Education and Tertiary Access at The University of the Sunshine Coast (USC), Australia. She is the Co-director of the Indigenous and Transcultural Research Centre and Chair of the Australian Community of Associate Deans Research in Education (cADRE), an Australian Council of Deans Education network. She is a historian who brings an innovative perspective to higher education research. Catherine has research projects on First Nations and transcultural doctoral education; academic identities; decolonising higher education and the transnational history of universities.

Lynn McAlpine is Professor Emerita at the University of Oxford, UK, and McGill University, Canada. She is internationally recognised for her research in the UK, Europe, and Canada about PhD and post-PhD career trajectories in and beyond academia.

Fanzhu Meng is pursuing her doctoral studies within the School of Education at the University of Glasgow. Her present research focuses on examining how international Chinese doctoral students successfully navigate the transitions within the UK education system. Her academic interests lie in the fields of psychological theories in education, critical and creative thinking, and innovative research methodologies.

Tai Peseta is an associate professor in the Academic Division, Western Sydney University (WSU), Australia. Following a five-year secondment to lead a university-wide curriculum transformation program, Tai is the Academic Lead, Student-Staff Partnership and heads a suite of educational initiatives—with the WSU Student Partner team—to grow partnership and co-inquiry across the university. She has published in the fields of doctoral supervision, academic development, critical university studies, and student-staff partnership.

Jenny Pizzica is now an independent scholar after professional and academic roles at the University of Notre Dame, Western Sydney University, University of Technology Sydney, and The University of Sydney.

Kelly Louise Preece is the Head of Academic Development and Enhancement at the University of Exeter. Over the course of her career, Kelly has worked in research, academic and professional services roles. Since joining the University of Exeter in 2015, she has developed a national reputation for her holistic and inclusive approach to supporting early career academics—including innovations in online and blended learning, sector leading writing support, and self-care for researchers. This expertise and innovative support was recognised when she became a National Teaching Fellow in 2022.

Jon Rainford is a staff tutor/lecturer in Education, Childhood, Youth and Sport at The Open University. He is also a visiting research fellow in Education at the University of Bath, UK. His doctorate focused on widening participation policy and practice and wider research focuses on student equity, transitions, and experiences. He is a co-convenor for the British Sociological Association Education Study group. He has recently co-edited *The Business of Widening Participation: Policy Practice and Culture* (Emerald) and has co-written *Thriving in Part-Time Doctoral Study* (Routledge) with Kay Guccione.

Yusuke Sakurai is an associate professor at the Center for Academic Practice and Resources at Hiroshima University in Japan. His main research interests include students' learning in international settings, pedagogy in higher education, and support for early career researchers. He has worked as an advisor for student peer support staff and the teaching assistant programme at Hiroshima University. Over the past 20 years, he has studied, taught, and conducted research in several countries, including Thailand, Australia, Malaysia, Egypt, Finland, and Japan (Tokyo and Hiroshima).

Gina Saliba After 12 years at Western Sydney University in a range of professional and academic roles related to strategic leadership of professional learning, Gina joined the Australian College of Applied Professions in April 2023 as an assistant dean responsible for academic leadership, overseeing the quality of learning and teaching on campus for all courses and units. She is completing her doctorate, titled 'Professional Learning in Unusual Contexts: Interrogating University Curriculum Groups'.

Fiona Salisbury joined Western Sydney University in early 2022 as Executive Director, Library Services, after a long career in library leadership at La Trobe University in Melbourne. She is responsible for the conceptualisation and implementation of several strategic library initiatives, including Library Reimagined and the Library of the Future. Fiona has published in numerous areas, among them, curriculum, student-staff partnership in the library, and information and research literacy. She is undertaking doctoral research that focuses on academic librarian's ideas of the university.

Isabelle Skakni is the head of the Researcher Career Development Unit at the University of Applied Sciences and Arts, Western Switzerland. Her research focuses on the training, career trajectories, and work experiences of doctoral researchers and PhD graduates, with a particular interest in how this population anticipates and prepares for its professional future, within and outside universities.

Kate Thomson is Senior Lecturer in Work Integrated Learning, at the Sydney School of Health Sciences with research expertise in professional learning. Her foundational research has centred on the professional learning of university teachers (i.e., clinical educators and academics) and her emerging interest is in developing future professionals through enhancing students' learning before, during, and after their work-integrated learning placement experiences. She has published over 50 articles, serves as an associate editor, *Higher Education Research & Development* journal, and is a senior fellow, Advance Higher Education.

Natthaphon Tripornchaisak is a postgraduate researcher in the School of Education and School of Modern Languages and Cultures investigating the notion of Intercultural Communication in the Thai–English interpreting industry. His research interests lie in the interdisciplinary bridge between intercultural communication and interpreting studies; intercultural mediation; language and culture; and creative research instruments. He is a full-time lecturer at the Department of English and Linguistics, Faculty of Liberal Arts, Thammasat University, Thailand. Natthaphon also works as a part-time Thai–English interpreter and translator.

Brittany Vermeulen (nee Hardiman) is the sustainability coordinator in Sustainability Education and Partnerships at Western Sydney University, and has received numerous local, national, and international accolades for her youth leadership in the area. Brittany is a doctoral researcher exploring learning environments for STEM students through short-term outbound mobility experiences.

Gina Wisker is a doctoral supervisor and senior lecturer/associate professor at the International Centre for Higher Education Management, University of Bath and Professor 11 at the University of the Arctic, Tromso where she contributes to doctoral supervisor development. Gina is a tutor on the OU's Realism and Fantasy A233 course. Gina runs workshops on supervisor development, doctoral writing and related research processes internationally including South Africa, Australia, Ireland, and Sweden. Her previous role was Professor of Higher Education and Contemporary Literature and Head of Centre for Learning and Teaching, University of Brighton where she is now Emeritus Professor (01.05.2006 to 31.4.2020). Prior to this, she had a similar role as head of the University Centre for Learning and Teaching at Anglia Ruskin University where she supervised doctoral students on a very large international part-time mid-career doctoral programme for 12 years. Gina is a practising teacher of contemporary, women's and postcolonial literature and writes about her work first to support research candidates and their supervisors and workshops and latterly in research-based articles and books, e.g. *The Good Supervisor* (2005, 2012) *Getting Published* (2015), *Getting Started with Supervision* (forthcoming 2023); website https://goodsupervisor.co.uk/. Gina also publishes in literature *Contemporary Women's Gothic Fictions* (2016) and *Contemporary Women's Ghost stories: Spectres, Revenants and Ghostly Returns* (2022). She is chief editor of SEDA journal *Innovations in Education and Teaching International* and dark fantasy online journal *Dissections.*

Zhihan Wu is a doctoral researcher in the School of Education, University of Glasgow. Her research looks at the shifting research self-efficacy throughout the pre-, mid-, and post-stages of Chinese international PGRs' research journeys. She is interested in human agency, learning autonomy, critical and creative thinking, metacognition, creative research methods, and research ethics.

Robyn Yucel is a senior lecturer and associate director, Learning Innovation in the Faculty of Science, Engineering and the Built Environment at Deakin University, Australia. Robyn conducts research in the philosophy of science for undergraduate science degrees. Her work in this area is strongly influenced by the meta-theory of critical realism. She also conducts research in higher education teaching and learning. Her most recent publication *Enhancing Powerful Knowledge in Undergraduate Education Science Curriculum for Social Good* was recently published in the journal *Teaching in Higher Education*.

Xun Zhang is a PhD student at the Graduate School of Human and Environmental Studies at Kyoto University. After receiving her master's degree from the Center for Japanese Studies at Beijing Foreign Studies University in China, she went on to study at Kyoto University, where her main research interests include language policy and history of foreign language education. She is working on her PhD dissertation on the history of foreign language teaching in Christian Colleges in nineteenth-century China.

List of Figures

Changing Career Pathways: Making Visible the Employment Destinations and Non-academic Contributions of Doctoral Scholars

Concluding Thoughts

List of Tables

Introduction

Dely Lazarte Elliot, Søren S. E. Bengtsen, and Kay Guccione

What This Book Aims to Offer

As someone interested in doctoral education, why might this book be relevant to you? How could this book be pertinent today? To address such questions, an ideal starting point is to discuss briefly selected but crucial doctoral concepts, i.e. *well-being*, *researcher independence*, *interdependence* and *the hidden curriculum*. In so doing, we raise two main questions: (1) How are these doctoral concepts perceived, understood and translated into practice? (2) What do these concepts mean for doctoral scholars' personal and professional development, supervisors' guidance and provision, researcher developers' and institutional leaders' approaches to supporting doctoral communities as well as for anyone who has a strong interest in

D. L. Elliot (✉) • K. Guccione
University of Glasgow, Glasgow, UK
e-mail: Dely.Elliot@glasgow.ac.uk; Kay.Guccione@glasgow.ac.uk

S. S. E. Bengtsen
Aarhus University, Aarhus, Denmark
e-mail: ssbe@edu.au.dk

D. L. Elliot et al. (eds.), *Developing Researcher Independence Through the Hidden Curriculum*,
https://doi.org/10.1007/978-3-031-42875-3_1

successful, meaningful and transformative doctoral education? It can be argued that how the first question is answered is strongly connected to how the second question is to be addressed.

We also aim to clarify competing concepts or widely held beliefs and then 'connect the dots' that underpin these various concepts, with a view to capturing and offering our readers a holistic view of the doctoral learning processes in the final chapter. Familiarity with existing, at times, differing conceptualisations of these ideas can assist in getting a sound grasp as these ideas are presented and illustrated in each chapter. What is more, this book is strongly pragmatic in nature. Our intention is to go beyond the theorised components of these concepts. Instead, it is to offer in each chapter practical demonstrations of how these concepts can be realised in various contexts.

Doctoral Well-being in the Spotlight

Within the fascinating world of doctoral education, shared issues typically confronting doctoral scholars that require urgent and ongoing attention are increasingly being recognised. This comes with a greater appreciation of how complex the doctoral landscape can be for its different cohorts—domestic, international, part-time or working PhD scholars. Arguably, this is something to be expected from anyone who embarks on the highest level of educational studies. It does not come as a surprise that one or a combination of its several dimensions (e.g. doctoral genre, research culture, academic working conditions) contributes to the discourse about doctoral scholars' mental health and well-being (Barry et al., 2018; Byrom et al., 2020; Elliot, 2023; McCray & Joseph-Richard, 2020; Metcalfe et al., 2018; Sverdlik et al., 2018). If left unaddressed, such well-being concerns may lead to or exacerbate other doctoral issues, e.g. lack of motivation and engagement, delay in progression, doctoral attrition or mental illness (Ayres, 2022; Devos et al., 2017; Elliot et al., 2023).

These studies exemplify and indicate the extent to which the doctoral population is plagued by this predicament. While Evans et al. (2018) suggest that the doctoral population tends to experience depression and anxiety about six times higher than the general population, Levecque et al. (2017) highlight the risk of developing a psychiatric disorder (e.g. depression) associated with doctoral academic working conditions. Accordingly, mental health concerns are far from being isolated experiences but could arise from a complex combination of personal psychological dispositions (e.g. impostor syndrome) even prior to the PhD, exposure to a toxic research culture and disquiet over the precarious working conditions awaiting post-doctoral scholars (Deconinck, 2015).

Mental health and well-being are not the main focus of this book, but since their immense influence to serve as a propeller or a barrier is recognised in each doctoral journey, you can expect well-being to be either explicitly or implicitly discussed in the chapters that follow.

Notably, such discussions contest the often implied discursive understanding of well-being found in policy and university strategy associating well-being with a clinical psychological discourse, thereby pathologising doctoral scholars before they have even begun their studies. This also often comes with a narrow view of well-being as being an individual, rather than a community and organisational phenomenon and linking well-being together with performativity agendas (Elliot et al., 2023; Petersen & Sarauw, 2023; Sarauw et al., 2023). On the contrary, many chapters in this book argue that well-being issues need to be tackled as social, knowledge-based and relational dimensions of the doctoral journey.

Researcher Independence and Interdependence: An Oxymoron?

Another crucial and related area that has generated attention in the doctoral literature concerns the connection between *independence* and *interdependence*. To this end, it is first worth addressing the widely held beliefs associated with researcher independence. Available literature on this topic signals the conflation between the term independence and working alone. While independence in conducting research denotes scholars' capacity to undertake research tasks with a high degree of autonomy, criticality, confidence and competence that equips them with a sense of research direction, we strongly argue that the process of achieving researcher independence can be attained through intentional pursuit of interdependence, thereby contesting the personal-social binary often unfruitful for doctoral formation (Bengtsen, 2016; Gardner, 2008).

As we see argued in several chapters in the book, it is through engaged discussions, collaborative pursuits or working together, or in other words, interdependent learning and critical thinking, that development naturally flourishes. In turn, it yields collective wisdom and more sustainable research communities, which offer scholars reciprocally meaningful learning for all involved. Further, observed interdependence among community members is widely recognised to offer social and emotional support (Cornér, 2020; Cornér et al., 2018). Beyond this support, such a community also often serves as a platform that invites learning of various forms—targeted and focused, incidental or random—via casual chats, informal brainstorming, exchange of ideas or peer mentoring, which tend to generate scholarly enrichment and cross-fertilisation of ideas (Elliot, 2023).

This is contrary to the widespread individualisation of researcher trajectories and careers encouraged by increased neo-liberal management of doctoral education, often focusing abstractly on the individual disconnected from the researcher, collegial, educational and institutional milieus and contexts within which the PhD is nested (McAlpine & Norton, 2006). Instead, we view researcher independence as a concept often catalysed and facilitated through collective effort, collaboration and community building (Cai et al., 2019; Elliot, 2023). Largely influenced by Vygotsky's social constructivism, this places learning as a collective effort among the community members (Daniels, 2016; Vygotsky, 1978) and more recent theory development within higher and doctoral education that focuses on ecological learning and doctoral ecologies (Barnett, 2018; Bengtsen, 2019). While seemingly oxymoronic, this view of researcher independence emanates from fostering interdependent practice (where individuals collectively influence each other's thoughts, actions and development—see Colman, 2015). Interdependence then facilitates deepening knowledge and subject expertise, broadening appreciation of concepts and ideas and expanding one's repertoire of skills.

Pursuit of researcher independence via interdependence stands in sharp contrast to the often misunderstood perception that developing independence means working alone, or rising above the crowd (i.e. the community). Instead, we argue that independence, thriving, creativity, criticality and originality in research rest on practising interdependence, relational trust-building and co-construction of knowledge found in balanced and sustainable researcher collectives (Guccione, 2016). This is based on the premise that researcher independence in the doctoral context is not only strongly conveyed but is developed through promoting interconnectedness and collective learning (e.g. reflective growth, navigation, leadership, enculturation, development of expertise and wealth of experience). Doing so is arguably even more crucial in the doctoral context where scholars generally operate within the constrictions of a pervasive culture of individualised working conditions.

So, What Might Researcher Independence Look Like in Practice?

For doctoral scholars, a sense of researcher independence might be characterised by combined acquisition of disciplinary knowledge, scholarly identity and a doctoral-level skillset—complemented by a set of dispositions and competencies, e.g. personal agency, creativity and innovation, critical and autonomous thinking, feedback literacy or use of an interdisciplinary approach (Åkerlind & McAlpine, 2017; Bastalich, 2017; Brodin, 2017; Guccione, 2016; Inouye & McAlpine, 2017; Johnson et al., 2000; Wisker et al., 2003).

Some doctoral scholars may already possess many of these characteristics even from the outset of their doctoral journey, and further development of researcher independence occurs during the doctoral process itself. Moreover, pursuit of researcher independence can pave the way for doctoral scholars' sense of ownership and direction over their own research priorities, capability to define and design their own research and conceptualise ways of knowledge generation. Perhaps, this can even lead to new collaborations and/or research grant applications—both during and after the PhD in continued academic or professional work environments. Since each doctoral study is distinct, formal and informal curricular lessons that reinforce researcher independence not only vary widely but manifest themselves differently.

Concurrently, there exist different views and understandings of when researcher independence begins to happen, what characterises this concept and what are the indicators that this quality has been attained (e.g. Åkerlind & McAlpine, 2017; Albertyn & Bennett, 2021; Elliot, 2022; Lovitts, 2005; Savva & Nygaard, 2021). While researcher independence is often strongly associated with doctoral learning and development, it is essential to acknowledge that there also exist several interpretations of this term. Therefore, a vital purpose of this book is to bring clarity to this multiplicity of interpretations in relation to researcher independence. More specifically, this book is expected to generate a comprehensive depiction of how formal and informal curricular elements can be harnessed specifically to foster researcher independence among doctoral scholars and how research environments and institutional culture may aid the formation of competent, independent doctoral researchers (Barnett et al., 2022; Overall et al., 2011).

Hidden Curriculum as a Vital Channel of Doctoral Pedagogies

Having considered key and contemporary challenges facing doctoral scholars, let us now turn to potential ways of addressing them. In so doing, it is vital to connect our discussion to this book's predecessor. In 2020, when our team published the book 'The Hidden Curriculum in Doctoral Education' (Elliot et al., 2020), the reception it received was indeed heartening. It also conveyed a strong indication of the 'thirst' to further explore this crucial topic. Likewise, there were questions raised concerning hidden curricular lessons' wider utility and applicability within (and beyond) the context of doctoral education—often with a hint of philosophical musings. Examples include: once found, is the hidden curriculum no longer 'hidden'? Can we disregard the hidden curriculum's previous negative connotations? Or is the scope of the hidden curriculum restricted to informal dimensions of learning? Equally, how do hidden pedagogies manifest themselves in different doctoral settings? How can we harness the potential of such doctoral pedagogies? All these questions suggest that there is much more to explore and that these questions are vital and deserve further attention. This interest and the increase of blog posts on this topic (e.g. see its accompanying website https://drhiddencurriculum.wordpress.com/) convinced our team that the time was ripe for a follow-up book on the hidden curriculum.

In our earlier book, we presented and discussed the negative connotations attached to the 'hidden curriculum' when it was first conceived (e.g. Gair & Mullins, 2001; Jackson, 1968; Martin, 1994). Originally contextualised in the school setting, its primary aim was to identify unintended messages and lessons conveyed through class activities, how learners are responded to, and even the classroom structure where learning took place (Jackson, 1968). Exposing the hidden curriculum then implies identifying and avoiding potential negative impacts of learning. Not discounting that the hidden curriculum could have a negative effect in any context (e.g. school, higher education), our team also argue how any unintended or incidental lessons and pedagogies arising from learning experience may equally bring about positive impact, and in turn, prove to be beneficial to learners. Consequently, this became the primary focus of our first book on the subject (Elliot et al., 2020). In this book, we continue to draw upon and build on our earlier book's conceptualisation of the hidden curriculum. At the end of what we regarded as a conceptual journey in writing this book, we proposed a more comprehensive definition of the hidden curriculum—a definition that we again employ for this edited collection.

> The hidden curriculum in doctoral education comprises all unofficial mechanisms of learning that take place within and outwith academia. Learning via the hidden curriculum is recognised as genuine pedagogical spaces or sites of learning that can extend pedagogical practices by offering support provision for learners' academic, personal, social and psychological needs. Whereas the starting point in the pursuit of the hidden curriculum tends to be driven by doctoral researchers' ownership of this personal process, the entire doctoral ecology recognises that there are key 'hidden curriculum agents' who are able to support, empower and enable doctoral researchers in creating learning pathways that are strategically intended to harness a tailored hidden curriculum based on personal needs and professional aspirations. (Elliot et al., 2020, pp. 130–131)

Our definition stressed the value of the entire doctoral ecology, which takes into account Barnett's (2018) concept of ecological university to explain how 'knowledge creation, learning and higher education curricula and institutions are typically embedded within a wider range of disciplinary, institutional, societal, political and existential contexts' (Elliot et al., 2020, p. 98). Moreover, it extends the notion of 'nested contexts' in doctoral education (McAlpine and Amundsen, 2016; Elliot et al., 2016) in order to highlight how doctoral learning is situated within a range of contexts that are inhabited by doctoral stakeholders or 'hidden curriculum agents'. While we acknowledged in the first book the complementary and supportive roles of these stakeholders, this very idea has inspired the focal point of this edited collection. Linking to the earlier discussion on interdependence, we aim to convey how personal and collective efforts among various communities in the doctoral nested systems can crucially improve the quality of the research culture. Such intention is reflected in the structure of this edited collection—affording various doctoral stakeholders a voice and a perspective on how they initiate, promote and support the development of researcher independence via the hidden curriculum.

Notably, exploring the positive dimensions of the hidden curriculum, particularly the strong pedagogical benefits embedded in it, is particularly apt in the doctoral learning context. While a lack of structure is core to the doctoral genre, it, nevertheless, comes with intellectual, social and psychological demands, as well as a high level of commitment. Therefore, increased understanding of the hidden curriculum is a means of maximising the tools and resources that doctoral scholars can meaningfully tap into. It is worth contending that although hidden curricular learning may come from both the formal and informal curriculum, simply prioritising institutional provision risks limiting doctoral scholars' potential for transformative development, researcher independence and career readiness.

Why Develop Researcher Independence Through the Hidden Curriculum?

In planning this new book, our team gravitated towards exploring two vital concepts in doctoral education—the hidden curriculum and researcher independence. For us, exploring them together is promising on two counts. Not only do we intend to bring a deeper understanding of these key concepts, but it is to examine their potential connections and, in turn, enable us to paint a fuller picture potentially to highlight their interwoven importance—both in theory and in practice. Apart from being research-based, and as discussed in the previous section, we want this book to capture various stakeholders' voices, their perspectives and their first-hand experience based on culturally and geographically diverse doctoral settings.

Employing an ecological approach in supporting researcher independence, we contend that it is critical to seek complementary perspectives from doctoral scholars, supervisors and mentors, researcher developers, institutional leaders and others (Barnett, 2018; Bengtsen, 2020; Elliot et al., 2020). Not only could they helpfully elucidate the 'fuzziness' surrounding the term researcher independence, but with authors coming from differing contexts, it has an added advantage of highlighting similarities, overlaps or differences to complement, enrich or challenge conceptual understanding and practices. The combined insights from the 45 book contributors who are equipped with disciplinary expertise and represent various geographical regions—from New Zealand, Australia, Japan, Hong Kong, Thailand, South Africa, Finland, Denmark, Sweden, Switzerland, Canada and United Kingdom—are also intended to offer a more comprehensive account and stimulate current thinking with a view to presenting examples of unconventional, yet effective doctoral pedagogical practices in support of doctoral scholars en route to becoming independent and competent researchers.

Having taken all these into account, we regard an edited collection to be the best way forward when probing researcher independence using the conceptual lens of the hidden curriculum. Through this edited collection, our shared aim is to elucidate potential theoretical links between these two important concepts while offering practical examples and demonstrating how the hidden curriculum may support the development of doctoral scholars' researcher independence. Each chapter is an attempt to highlight, discuss and exemplify the instrumental and formational roles played by the hidden curriculum in promoting or facilitating researcher independence. Our secondary aim is that in addressing these doctoral concepts, not only will we contribute to the limited scholarly resources on the hidden curriculum, but we can also stimulate conversation and debate concerning its theoretical and pragmatic importance in reinforcing doctoral scholars' transformation into independent researchers. In sum, each chapter is filled with conceptual and practical insights from different perspectives and contexts giving this book a strong applied focus as we (editors and all authors) collectively examine if and how the hidden curriculum may serve as a channel for fostering or strengthening researcher independence.

How Is This Book Designed to Achieve Its Core Aim?

In executing our book's primary aim, serious consideration was given to various factors:

- The book conveys views from doctoral scholars, doctoral supervisors, researcher developers, institutional leaders and other stakeholders outside academia concerning the hidden curriculum in the doctoral context. We capitalise on the research-based perspectives and first-hand examples of these key players in doctoral education to crystallise what developing researcher independence entails.
- Expert contributions from different geographical regions are intended to offer complementary insights and enrich current understanding. Likewise, they are expected to raise contestations and tensions, challenge current understanding or offer an alternative

appreciation (e.g. researcher independence) where differences of perspective are put forward.

- Each chapter typically features a conceptual model, experience-based observations and reflections, and/or complemented by an empirical study to demonstrate how the hidden curriculum may facilitate and sustain the development of researcher independence, including preparation for the post-PhD stage.
- With the book's pragmatic focus, we ensure that a range of voices from differing doctoral groups are included, e.g. domestic, international, part-time or working PhD scholars as well as of early career scholars.
- We strategically curated each chapter of this book, initially from many authors who responded to a request for a 1000-word blog post. Doing so has established a firm basis for capturing the authentic voices of the international research community.
- Finally, each chapter intends to clarify the academic, institutional and pedagogical 'fuzziness' surrounding conceptualisation and development of researcher independence using hidden curricular pedagogies. It attempts to offer examples of what 'striking a balance' may mean in terms of what might be the required level of autonomy and academic support, accountability and boundaries when supporting doctoral scholars' personal and professional development (and/or their supporters, e.g. supervisors, researcher developers) (Benmore, 2016; Overall et al., 2011; Wisker et al., 2003).

The Division of the Book

This book is strategically divided into five parts to represent the wealth of perspectives from key stakeholders in doctoral education. Twenty-two chapters have been grouped according to five doctoral stakeholder categories in order to:

(a) contribute to the meagre scholarly resources on the hidden curriculum within the context of doctoral education with its strong emphasis on hidden curriculum's pedagogical benefits;
(b) elucidate the interconnection between hidden curriculum and researcher independence to achieve a better appreciation and a more holistic view of the doctoral process; and

(c) offer hidden curriculum-focused theoretical, conceptual and practical contributions containing further reflection and scrutiny of existing literature, in-context observations, first-hand experiences or empirical studies from a number of book chapter contributors across the globe.

The Conclusion chapter will draw upon all the chapters in order to synthesise the multiplicity of ideas and messages highlighted throughout the book—from 'food for thought' through to alternative approaches or consideration of other doctoral pedagogical practices. Primarily, it is to support our contention that the hidden curriculum plays a central role in developing doctoral scholars' researcher independence. Finally, we will discuss and elaborate on what the implications are for all doctoral stakeholders—doctoral scholars themselves, supervisors, researcher developers and institutional leaders.

Each part of the book is briefly presented below.

Part I: Insights from Doctoral Scholars

These five chapters comprise a combination of ethnographic accounts from doctoral scholars as they reflect on: (a) identifying retrospectively 'hidden curriculum agents' and their long-term influence on scholarly independence; (b) mapping out the learning opportunities offered by the hidden curriculum specifically in an international doctoral setting; (c) appraising the developmental value of peer mentoring; (d) critically examining hidden curricular lessons when returning from overseas PhD study; and (e) exercising interdependence and developing researcher competence via participation in a Journal Club.

Part II: Insights from Doctoral Supervisors

In the next four chapters, discussion will revolve around how doctoral supervisors may strategically embed and harness the hidden curriculum for doctoral scholars' benefit via: (a) capitalising on 'Fridaying' and other supervisor and doctoral interactions as 'forms of dynamic developmental dialogues' to demystify doctoral processes leading to successful researcher independence; (b) unlocking and stimulating doctoral scholars' independence, interdependence and creativity by differentiating 'creative supervising' and 'supervising for creativity' as pedagogical strategies; (c) examining the potential role of metacognition not only in navigating the doctoral experience but in strengthening a sense of researcher independence; and (d) entering into the #thesisthinkers project—a negotiated partnership with doctoral scholars involving 'co-creation' of their own curriculum.

Part III: Insights from Researcher Developers

The five chapters that follow investigate further contributory concepts, practices and pedagogies in relation to fostering researcher independence. These chapters exemplify how researcher independence can be pinned down, and, in turn, applied via a deeper appreciation of the formal and the hidden curricula as well as the interaction between them. These chapters specifically consider: (a) development and evaluation of pedagogical practices designed to support scholars at all stages of the doctoral journey by tapping into both formal and informal ways of learning and multiple ways of doing; (b) creative use of humour in conveying the desirability of understanding PhD norms and expectations, connecting doctoral standards and values through supervisory practices and negotiating boundaries—all with a view to building doctoral scholars' sense of independence; (c) instilling confidence in writing and enhancing well-being via participation in doctoral writing groups, in which doctoral scholars serve as a valuable resource to foster each other's success in the research environment; (d) conceptualising doctoral intelligence framed with the four domains to guide dynamic doctoral researcher development, i.e. 'knowing', 'doing', 'thinking' and 'willing' mindsets; and (e) developing effective or multiple support villages for part-time doctoral scholars who typically have limited interactions.

Part IV: Insights from Institutional Leaders

Drawing upon the perspectives of those whose remit involves Doctoral and Research Leadership, four chapters are designated to raising both crucial and timely concepts and issues for consideration by institutional leaders, given today's doctoral education challenges informing overall practice. This section covers: (a) the value of skilled coaching, mentoring, good quality peer and supervisor conversations at the centre of managing doctoral learning experience; (b) recognising doctoral education both as a key site for knowledge production and as an avenue to reclaim, revive and extend indigenous and transcultural knowledge to harness 'hidden reservoirs of knowledge and agency'; (c) the juxtaposition (and intriguing thought) behind the ideas of leadership for doctoral education, hidden curriculum and researcher independence; and (d) enactment of practices within institutional leadership in doctoral education based upon the interplay between opportunity structures in the institution.

Part V: Insights on Doctoral Education Beyond Academia

The last four chapters illuminate the value of researcher independence following doctoral completion and more broadly, i.e. beyond academia. In these chapters, the authors clarify and discuss more deeply: (a) the 'extra' in extracurricular argued to be integral in enriching the doctoral experience—with featured examples of such learning; (b) the misaligned, hidden or ambiguous 'meanings' of researcher independence and their implications for doctoral recruitment, development, provision, supervisory practice and assessment; (c) the transition from doctoral study to post-PhD work, particularly to careers outside academia, where a mismatch between prior experience and organisational culture and practices require negotiation and adjustment; and (d) an analysis of employment data to inform and empower doctoral scholars' career planning endeavours, to value doctoral skills beyond academia and to strengthen their sense of agency to increase their readiness for the next phase after their PhD.

References

Åkerlind, G., & McAlpine, L. (2017). Supervising doctoral students: Variation in purpose and pedagogy. *Studies in Higher Education, 42*(9), 1686–1698. https://doi.org/10.1080/03075079.2015.1118031

Albertyn, R., & Bennett, K. (2021). Containing and harnessing uncertainty during postgraduate research supervision. *Higher Education Research & Development, 40*(4), 661–675. https://doi.org/10.1080/07294360.2020.1775559

Ayres, Z. J. (2022). *Managing your mental health during your PhD: A survival guide*. Springer.

Barnett, R. (2018). *The ecological university. A feasible utopia*. Routledge.

Barnett, R., Bengtsen, S., & Nørgård, R. T. (2022). *Culture and the university. Education, ecology, design*. Bloomsbury.

Barry, K., Woods, M., Warnecke, E., Stirling, C., & Martin, A. (2018). Psychological health of doctoral candidates, study-related challenges and perceived performance. *Higher Education Research and Development, 37*(3), 468–483. https://doi.org/10.1080/07294360.2018.1425979

Bastalich, W. (2017). Content and context in knowledge production: A critical review of doctoral supervision literature. *Studies in Higher Education, 42*(7), 1145–1157. https://doi.org/10.1080/03075079.2015.1079702

Bengtsen, S. (2016). *Doctoral supervision. Organization and dialogue.* Aarhus University Press.

Bengtsen, S. (2019). Building doctoral ecologies and ecological curricula. The sprawling spaces for learning in researcher education. In Jackson, N. & Barnett, R. (Eds.). *Learning ecologies* (pp. 146–159). Routledge.

Bengtsen, S. (2020). Building doctoral ecologies and ecological curricula. Sprawling spaces for learning in researcher education. In R. Barnett, & N. Jackson (Eds.). *Ecologies for learning and practice. Emerging ideas, sightings, and possibilities* (pp. 147–159). Routledge.

Benmore, A. (2016). Boundary management in doctoral supervision: How supervisors negotiate roles and role transitions throughout the supervisory journey. *Studies in Higher Education, 41*(7), 1251–1264. https://doi.org/10.1080/03075079.2014.967203

Brodin, E. M. (2017). The stifling silence around scholarly creativity in doctoral education: Experiences of students and supervisors in four disciplines. *Higher Education, 75*, 655–673. https://doi.org/10.1007/s10734-017-0168-3

Byrom, N. C., Dinu, L., Kirkman, A., & Hughes, G. (2020). Predicting stress and mental wellbeing among doctoral researchers. *Journal of Mental Health, 1-9.* https://doi.org/10.1080/09638237.2020.1818196

Cai, L., Dangeni, D., Elliot, D. L., He, R., Liu, J., Makara, K. A., Pacheco, E.-M., Shih, H.-Y., Wang, W., & Zhang, J. (2019). A conceptual enquiry into communities of practice as praxis in international doctoral education. *Journal of Praxis in Higher Education, 1*(1), 11–36.

Colman, A. M. (2015). *Oxford dictionary of psychology.* Oxford University Press.

Cornér, S. (2020). *The socially-embedded support system in doctoral education*, PhD Dissertation, University of Helsinki.

Cornér, S., Pyhältö, K., Peltonen, J., & Bengtsen, S. (2018). Similar or different? Researcher community and supervisory support experiences among Danish and Finnish social sciences and humanities PhD students, *Studies in Graduate and Postdoctoral Education.* https://doi.org/10.1108/SGPE-D-18-00003

Daniels, H. (2016). *Vygostky and pedagogy.* Routledge.

Deconinck, K. (2015). Trust me, I'm a doctor: A PhD survival guide. *The Journal of Economic Education, 46*, 360–375.

Devos, C., Boudrenghien, G., Van der Linden, N., Azzi, A., Frenay, M., Galand, B., & Klein, O. (2017). Doctoral students' experiences leading to completion or attrition: a matter of sense, progress and distress. *European Journal of Psychology of Education, 32*(1), 61–77. https://doi.org/10.1007/s10212-016-0290-0

Elliot, D. L. (2022). A 'doctoral compass': Strategic reflection, self-assessment and recalibration for navigating the 'twin' doctoral journey. *Studies in Higher Education, 47*(8), 1652–1665. https://doi.org/10.1080/03075079.2021.1946033

Elliot, D. L. (2023). *Navigating your international doctoral experience (and beyond)*. Routledge. ISBN 9781032220505

Elliot, D. L., Baumfield, V., Reid, K., & Makara, K. A. (2016). Hidden treasure: Successful international doctoral students who found and harnessed the hidden curriculum. *Oxford Review of Education, 42*(6), 733–748.

Elliot, D. L., Bengtsen, S. S. E., Guccione, K., & Kobayashi, S. (2020). *The hidden curriculum in doctoral education*. Palgrave Pivot. https://doi.org/10.1007/978-3-030-41497-9

Elliot, D. L., Ohlsen, S., Guccione, K., Daley, R. A., and Blackmore, C. (2023, in press). The PhD is a personal individual struggle … but you don't have to struggle alone: Supervisors' perspectives of international scholars' well-being. In: Edwards, M., Martin, A., and Ashkanasy, N. (eds.) *Handbook of academic mental health*. Edward Elgar Publishing.

Evans, T. M., Bira, L., Gastelum, J. B., Weiss, L. T., & Vanderford, N. L. (2018). Evidence for a mental health crisis in graduate education. *Nature Biotechnology, 36*(3), 282–284. https://doi.org/10.1038/nbt.4089

Gair, M., & Mullins, G. (2001). Hiding in plain sight. In E. Margolis (Ed.), *The hidden curriculum in higher education* (pp. 21–41). Routledge.

Gardner, S. K. (2008). "What's too much and what's too little?": The process of becoming an independent researcher in doctoral education. *The Journal of Higher Education, 79*(3), 326–350.

Guccione, K. (2016). *More than lucky? Exploring self-leadership in the development and articulation of research independence*. L. F. f. H. Education.

Inouye, K. S., & McAlpine, L. (2017). Developing scholarly identity: Variation in agentive responses to supervisor feedback. *Journal of University Teaching & Learning Practice, 14*(2), 1–19.

Jackson, P. (1968). *Life in classrooms*. Holt, Rinehart and Winston Inc.

Johnson, L., Lee, A., & Green, B. (2000). The PhD and the autonomous self: Gender, rationality and postgraduate pedagogy. *Studies in Higher Education, 25*(2), 135–147. https://doi.org/10.1080/713696141

Levecque, K., Anseel, F., De Beuckelaer, A., Van der Heyden, J., & Gisle, L. (2017). Work organization and mental health problems in PhD students. *Research Policy, 46*(4), 868–879. https://doi.org/10.1016/j.respol.2017.02.008

Lovitts, B. E. (2005). Being a good course-taker is not enough: A theoretical perspective on the transition to independent research. *Studies in Higher Education, 30*(2), 137–154. https://doi.org/10.1080/03075070500043093

Martin, J. R. (1994). What should we do with a hidden curriculum when we find one? In *Changing the educational landscape: Philosophy, women and curriculum* (pp. 154–169). Routledge.

McAlpine, L., & Amundsen, C. (2016). *Post-PhD career trajectories. Intentions, decision-making and life aspirations.* Palgrave Macmillan.

McAlpine, L., & Norton, J. (2006). Reframing our approach to doctoral programs: A learning perspective. *Higher Education Research and Development, 25*(1), 3–17.

McCray, J., & Joseph-Richard, P. (2020). Towards a model of resilience protection: Factors influencing doctoral completion. *Higher Education, 80*, 679–699. https://doi.org/10.1007/s10734-020-00507-4

Metcalfe, J., Wilson, S., & Levecque, K. (2018). *Exploring wellbeing and mental health and associated support services for postgraduate researchers.* Vitae in partnership with the Institute for Employment Studies and the University of Ghent.

Overall, N. C., Deane, K. L., & Peterson, E. R. (2011). Promoting doctoral students' research self-efficacy: combining academic guidance with autonomy support. *Higher Education Research & Development, 30*(6), 791–805. https://doi.org/10.1080/07294360.2010.535508

Petersen, E. B., & Sarauw, L. L. (2023). Psychometrics as performance indicator: New modes of governance of universities? *International Journal of Qualitative Studies in Education.* https://doi.org/10.1080/09518398.2023.2181460

Sarauw, L. L., Bengtsen, S., & Filippakou, O. (2023). The psychological turn in higher education and the new taxonomy of attitudes and emotions: Denmark as a case study. *Policy Futures in Education.* https://doi.org/10.1177/14782103231173017

Savva, M., & Nygaard, L. P. (2021). *Becoming a scholar. Cross-cultural reflections on identity and agency in an education doctorate* (M. Savva & L. P. Nygaard, Eds.). UCL Press.

Sverdlik, A., Hall, N. C., McAlpine, L., & Hubbard, K. (2018). The PhD experience: A review of the factors influencing doctoral students' completion, achievement, and well-being. *International Journal of Doctoral Studies, 13*, 27.

Vygotsky, L. (1978). Mind in society: The development of higher psychological processes. .

Wisker, G., Robinson, G., Trafford, V., Warnes, M., & Creighton, E. (2003). From supervisory dialogues to successful PhDs: Strategies supporting and enabling the learning conversations of staff and students at postgraduate level. *Teaching in Higher Education, 8*(3), 383–397. https://doi.org/10.1080/13562510309400

PART I

Insights from Doctoral Scholars

Through and Towards an interdisciplinary Research Community: Navigating Academia as a Lone Doctoral Scholar

Melina Aarnikoivu

Introduction: An Accidental Doctoral Scholar

I became a doctoral scholar by a complete accident. Of course, the doctoral study position did not simply drop down from the sky; to acquire it, I had to create a solid research plan, discuss my ideas with potential supervisors, and finally become accepted to a doctoral programme at the Centre for Applied Language Studies, University of Jyväskylä. However, what led me to do all this was a series of unplanned, serendipitous events, which were *not* on my to-do list when I finished my Master's degree in 2012. At the time, in fact, I had said to myself that I would never write a single academic text in my life again. Obviously, I was wrong.

If I could characterise my work in academia with one word, it would be: coincidences. Still, after two and a half years since obtaining my PhD, I do

M. Aarnikoivu (✉)
Finnish Institute for Educational Research, University of Jyväskylä, Jyväskylä, Finland

D. L. Elliot et al. (eds.), *Developing Researcher Independence Through the Hidden Curriculum*,
https://doi.org/10.1007/978-3-031-42875-3_2

not have a career plan or a clear idea what I want to do "when I grow up". Therefore, it is challenging for me to advise or provide tips to potential doctoral scholars. How can I tell others how to plan their doctoral studies when mine were very messy: working outside of a research group, from abroad, first part-time and later with several short, separate funding periods? Who am I to mentor anyone, knowing that all doctoral journeys are different, depending on one's background, the country or the university where they are studying, the discipline(s), the surrounding academic community, and the sudden surprise opportunities that we have no way of knowing about before they are right there in front of us.

What I *can* do, however, is to look back at all the happenstances of my doctoral journey and reflect on *what was actually going on*. What were the key opportunities I spotted and took? In this chapter, I provide an autoethnographic (Ellis et al., 2011) account of two *key incidents* (Kroon & Strum, 2007) which ended up shaping my doctoral studies in a way I did not perhaps realise at the time, but retrospectively examined had a long-term influence on how my future research and collaboration would look like. As Elliot et al. (2020) pointed out, however, *learning* that happens within hidden curriculum might often seem accidental or unintentional but should not be underestimated because of that. Instead, this type of learning can be genuine and valuable, and, therefore, as important as the one taking place within formal curriculum.

Given that my research topic as well as my *learning space* during my doctoral studies was highly interdisciplinary, I have built the chapter around the concept of *the world of opportunities* by Brodin and Avery (2020), who examined early-career researchers' development of researcher independence in multidisciplinary learning environments. In their article, Brodin and Avery (2020) presented two ways of developing researcher independence: *away from* the epistemic/social community or *through and toward* the community (Brodin & Avery, 2020, p. 420). According to them, what determines the outcome between these two trajectories is (1) the quality of the social interactions but also (2) one's temporal (career-stage) and spatial (geography, epistemology) position. For example, if a doctoral scholar is supervised by a highly mono-disciplinary supervisor, it is possible that they will not encounter viable opportunities for interdisciplinary engagement in their learning space. Alternatively, a doctoral scholar might feel lack of acceptance (of their knowledge) in the interdisciplinary environment that they are in. In both these cases, the interdisciplinary community might feel *alien* or become *avoided* altogether. By contrast, doctoral scholars who have chances to engage in interdisciplinary learning spaces also have the most optimal conditions to create collaborative networks, as well as to develop their researcher independence (Brodin & Avery, 2020).To illustrate how I developed my researcher independence *through and towards* the community of higher education researchers, I first present two key incidents which significantly contributed to me gradually becoming an independent scholar. In both incidents, the hidden curriculum and *hidden curriculum agents*—other doctoral scholars, colleagues, and supervisors supporting my doctoral journey (Elliot et al., 2020)—had a crucial role in my doctoral learning. After presenting the key incidents, I will show how the positive learning experiences during my doctoral studies led to several research networks and collaborations now that I am a postdoctoral researcher, before concluding the chapter. By doing this, I want to offer valuable insights to other early-career researchers engaged in interdisciplinary research, particularly the ones working outside of a research group.

Key Incident 1: Organising an International Conference

In the spring of 2017, my PhD funding was about to come to an end. By that time, I had been working on my doctoral project for a year and a half; first part-time alongside a non-research job at the university, and afterwards on two short grants from my department. Fortunately, I did not have to abandon my goal of becoming a doctor, as I was then contacted by the institution where I had worked previously, inquiring if I would be willing to accept a six-month contract to work on a research project on multidisciplinary peer-mentoring and to organise an international higher education research conference. Having no other funding in sight, I accepted the job offer, even though it meant less time for working on my thesis.

Both the research project and the conference organisation turned out to be highly valuable experiences not only in terms of doctoral learning but also in developing researcher independence towards the higher education research community. The tacit knowledge I acquired during those six months, especially prior and during the conference, was my first introduction to the "messy world" of higher education research. As many scholars have concluded, *higher education research* is a scattered field and difficult to define (e.g. Clegg, 2012; Macfarlane & Grant, 2012; Tight, 2020). This is because those studying higher education represent several different "background disciplines", such as education, sociology, economics, or history, meaning that whenever higher education researchers come together—to publish at a journal or present at a conference, for example—the epistemological understanding of those involved might highly differ from each other. Negotiating and reflecting those differences could be considered vital for those who want to work their way *towards* the higher education research community. Otherwise, that community might remain alien or become avoided, in Brodin and Avery's (2020) terms.

The reason why understanding this mélange of researchers and their backgrounds was so important to me at the time was because I would also participate in the conference as a doctoral scholar, presenting my preliminary research on doctoral education. In other words, this versatile crowd of higher education researchers would be my audience and potential future collaborators. Thus, knowing who would be in the conference, what they had researched before, who they were doing research with, and whether or not they were approachable for an early-career researcher were some of the issues that were important but that I could not possibly know without help.

I was organising the conference with a colleague whom I had known for five years at that point and who, like me, was doing their PhD in higher education research. Unlike me, however, they had already attended several international higher education research conferences, written articles with several scholars around the world, and worked on different projects for several years. From this colleague (and friend), I would come to learn a great deal in those few months that we were planning the event. The learning would happen in the most mundane of situations: while creating the conference programme, planning who should present with whom in the same session, responding to participants' emails, or creating nametags for them.

While the formal outcome of the conference—the fact that it happened, as well as my presentation in it—is what stands in my CV, it is not what was the most helpful for me as a doctoral scholar at the time. Instead, it were those boring, often also quite hectic moments in the midst of the organisation process that would serve as valuable *mini-learning opportunities* (Elliot et al., 2020, p. 6). By doing this organisational work with my colleague, I gradually began to get a preliminary idea of the higher education researcher community and whether or not I wanted to be part of it in the future. Given that my experiences of mono-disciplinary conferences in linguistics had not been equally positive, I had a feeling that I had found "my crowd".

Key Incident 2: ECHER

About a year after the conference, I received an email from the same colleague I had organised the conference with. They explained that there was a small group of scholars who were thinking of "reviving" the network for Early-Career Higher Education Researchers (ECHER[1]), which had been established in 2011 but had since become dormant. My colleague had recommended me for the other group members to become involved due to my interest in doctoral education and my writing skills. The opportunity sounded fascinating, so I said yes and met with the group.

A few months later we launched the ECHER blog, which celebrated its fourth anniversary in December 2022. During these four years, ECHER has published dozens of blog posts on higher education research, written mostly by early-career scholars around the world. We have also gathered resources on various topics, such as academic writing and higher education research as a field, and interviewed editors of higher education research journals about writing and publishing.[2]

Again, however, it is not the visible outcomes, such as the blog posts, that have contributed most to my (post)doctoral learning. Even though creating those texts has been interesting and taught me a great deal about writing and editing, it is the informal, hidden communication "behind the scenes" that has been the most beneficial in terms of becoming part of a higher education research community. Within the past four years, I have had hundreds of hours of informal chatter and email exchanges with people around the world that I know because of ECHER. In these chats, I have learned about writing and publishing, events, interesting scholars and their work, differences between different higher education systems, and generally about how to navigate the 21st academia as an early-career higher education scholar. In other words, for me, ECHER has become a network of each other's hidden curriculum agents, sharing the tacit

[1] https://echer.org/

[2] A colleague and I wrote about this journey and our community development in 2021 (see Brankovic & Aarnikoivu, 2021).

knowledge we have received throughout our academic journeys in a variety of disciplines.

Had I said "no" to ECHER back in 2018, my research network would most likely be much smaller now. Surely, there might have been other, alternative networks for me to join but ECHER has given me a unique space to grow as a researcher, as being its co-coordinator has allowed me to do more or less what I want. Having such freedom as a doctoral scholar is uncommon. Finding like-minded early-career scholars willing (or able) to do a great deal of voluntary work for an online community is also uncommon. Moreover, I would most likely not have gone and taught in Siberia in 2020–2021, met many of my current colleagues, or even do research on academic writing because I would not have met some of the key people who, in one way or the other, contributed to the path that I am on right now. Finally, and most importantly, I would not have been able to see all the fruitful discussions that can take place when early-career higher education researchers from all continents come together to discuss the challenges of the twenty-first-century higher education.

Transitioning to Being a Postdoctoral Researcher: Who to Say "Yes" or "No" To?

Moving on to the postdoctoral stage, I have kept working outside of an established research group. While it is sometimes lonesome not to have a group with which to meet regularly to discuss a specific research topic, I have constantly been able to create new joint projects with researchers around the world. This has been essential, as I have kept working at distance from abroad, not having a regular campus access.

What has been different within the past two years compared to my doctoral studies is that my research network has quickly grown immensely. These new collaborators did not appear from nowhere: I knew them because of something I already did during my PhD—through a conference or ECHER, for example. And, like the two key incidents above, also my current networks and collaborations are largely a result of an accident—being in the right place at the right time, and expressing interest to the right people. In Table 1, I have listed my current research networks and collaborations, when they began, how I became involved, and what their explicit or hidden outcomes are.

Table 1 My postdoctoral collaborations or networks and their origin

Collaboration or network	*Discipline/field/ theme*	*Start (and end)*	*How I became involved*	*Outcomes*
Network of Early-Career Higher Education Researchers (ECHER); ~400 members from around the world	Higher education research (multiple disciplinary backgrounds)	Fall 2018–	A colleague told another scholar that I have the skills needed to edit a blog	Network coordination, an academic blog, online meetings, further networking, a published paper
Journal of Praxis in Higher Education (JPHE): An academic journal with an editorial team of ~15 scholars from several different countries	Praxis in higher education	2019–2023	I volunteered to become part of the editorial team at a symposium	Publishing an OA journal in higher education research
Postdoctoral projects on academic writing (four collaborators)	Academic writing and writing support	2020–	I was contacted by different co-authors by email. Some of them knew me from before, some of them did not	Three published papers (currently in progress); future ideas for collaboration
A peer-mentoring group consisting of early- and mid-career women scholars	Doctoral education	Fall 2021	I knew a scholar who had joined the group earlier and later invited me to become part of the group as well	Regular support, potentially an open network/ platform for researchers around the world
A COST Action (ReMO)	Researcher mental health	Fall 2021	My article on peer-mentoring was spotted by the COST Action Grant Manager	Several initiatives around the topic of researcher mental health and wellbeing

(*continued*)

Table 1 (continued)

Collaboration or network	*Discipline/field/ theme*	*Start (and end)*	*How I became involved*	*Outcomes*
Alternative Internationalisms group consisting of several scholars around the world	Internationalisation of higher education	Spring 2022	I was contacted by my PhD supervisor, asking if I was interested in joining a writing project	A published article-to-be in a special issue; regular online meetings

Looking at Table 1, and particularly the column *How I became involved*, it becomes clear that most collaborations are a result of someone knowing someone, being in the right place at the right time, or someone reading or hearing about your earlier work. In fact, I might claim that none of the items listed in the table resulted from me consciously pursuing those opportunities. Instead, they came to be by accident and by seizing the opportunity when it presented itself because I felt it would be something fun and interesting at the time.

The *world of opportunities* has, in fact, become so immense that I have had to turn some offers down, no matter how interesting they have been. This is where the skill of saying "no" has been incredibly important, as most opportunities are not typically funded, at least in the beginning. Which opportunities to take is, however, not an easy question, and sometimes there are attempts to collaborate which simply end up wasting time. While we can never get those hours back, what we can do is learn to recognise what makes a collaboration successful. For me, it is not the "prestige"—publishing in "top journals" with those from "top universities". Nor is it getting a long list of good-looking CV items or ticking boxes of "this is what I should do to advance in my career". Instead, it is working with people whose ways of working, research interests, and research values align with my own—people who work in higher education to make it more accessible, just, and equitable.

Conclusion

In this chapter, I have looked back at two key incidents that ended up shaping my research networks and collaboration in unexpected, yet positive ways. In these two interdisciplinary learning spaces—an international higher education conference and ECHER—I could develop my researcher independence *through and towards* the community of higher education researchers, instead of away from it. As a result, the world of higher education research became familiar and comfortable to me, rather than alien or avoidable. This, in turn, helped me to form further research networks and collaborations within higher education research after obtaining my PhD.

As Brodin and Avery (2020) pointed out, for the work towards the community to be successful, (1) the quality of the social interactions and (2) the scholar's temporal/spatial position matter. Looking at (1), there were two hidden curriculum agents or group of agents who shared their tacit knowledge on the community, helping me to navigate it. Without these interactions, it would have been more difficult for me to understand the interdisciplinary field of higher education research as a doctoral scholar with a background in applied linguistics. Moreover, I argue that it would have been more difficult for me to learn to recognise which opportunities in Table 1 to take, and which opportunities to pass. In itself, such skill is crucial in becoming an independent researcher: instead of doing what other scholars seem to be doing—or telling others to do—assessing what types of opportunities are suitable for one's own researcher and career development is a must not only as a doctoral scholar but also later in one's career.

Looking at (2), not only was I a novice researcher trying to make sense of what my researcher identity is but also someone fairly detached from the physical environment where both formal and informal doctoral learning typically took place—at the campus and its meeting rooms, offices, corridors, and coffee rooms. It was not until COVID-19 that this would drastically change, but for me, having these interdisciplinary learning spaces was crucial as a doctoral scholar who was primarily working at distance during a time when most other people were not. Without such spaces, it is likely that I would have become alienated from the higher education researcher community or actively began avoiding it by exploring other types of career alternatives after finishing my PhD.

Doing a PhD is difficult in many ways, and it is difficult in different ways, depending on what your background is, what you are studying, and where. However, what is shared by all researchers is that we cannot work alone—we need others, and we need communities for doing our work. (Learning via communities is explored further in Makara et al.'s and Rainford's chapters.) Therefore, the role of both hidden curriculum and hidden curriculum agents, whether doctoral scholars themselves, their supervisors, research developers, and institutions (Elliot et al., 2020), is of crucial importance when ensuring that all doctoral scholars have the necessary support to work *through and towards* a community, whichever community that is.

Declarations of Conflicts of Interest No conflicts of interest.

Acknowledgements I would like to thank my two most important "hidden curriculum agents", Jelena Brankovic and Taru Siekkinen, who have offered invaluable support and encouragement on my interdisciplinary academic path.

References

Brankovic, J., & Aarnikoivu, M. (2021). The making of early career higher education researchers. *Internationalisation of Higher Education, 2021*(1), 97–109. https://doi.org/10.36197/INT.1-2021.05

Brodin, E. M., & Avery, H. (2020). Cross-disciplinary collaboration and scholarly independence in multidisciplinary learning environments at doctoral level and beyond. *Minerva, 58*(3), 409–433. https://doi.org/10.1007/s11024-020-09397-3

Clegg, S. (2012). Conceptualising higher education research and/or academic development as 'fields': A critical analysis. *Higher Education Research and Development, 31*(5), 667–678.

Elliot, D. L., Bengtsen, S., Guccione, K., & Kobayashi, S. (2020). *The hidden curriculum in doctoral education.* Palgrave Pivot.

Ellis, C., Adams, T., & Bochner, A. P. (2011). Autoethnography: An overview. *Forum Qualitative Sozialforschung/Forum: Qualitative Social Research, 12*(1), 273–290.

Kroon, S., & Strum, J. (2007). International comparative case study research in education: Key incident analysis and international triangulation. In W. Herrlitz, S. Ongstad, & P. van de Ven (Eds.), *Research on mother tongue education in a comparative international perspective: Theoretical and methodological issues* (pp. 99–118). Rodopi.

Macfarlane, B., & Grant, B. (2012). The growth of higher education studies: From forerunners to pathtakers. *Higher Education Research & Development, 31*(5), 621–624.

Tight, M. (2020). Higher education: Discipline or field of study? *Tertiary Education and Management, 26*(4), 415–428. https://doi.org/10.1007/s11233-020-09060-2

Mapping the Learning Opportunities of the Hidden Curriculum for International Doctoral Scholars in Japan

Yusuke Sakurai
Equal Contribution Jiawen Han, and Xun Zhang

Introduction

The increase in doctoral scholars from abroad—international doctoral scholars (IDSs)—deserves attention to develop national and institutional initiatives towards a global knowledge society in Japan. To explore these

Jiawen Han and Xun Zhang have contributed equally for this chapter.

Y. Sakurai (✉)
Hiroshima University, Hiroshima, Japan
e-mail: sakurai@hiroshima-u.ac.jp

J. Han
Tianjin University of Technology, Tianjin, China
e-mail: hanjiawen@email.tjut.edu.cn

X. Zhang
Kyoto University, Kyoto, Japan
e-mail: zhang.xun.63m@st.kyoto-u.ac.jp

D. L. Elliot et al. (eds.), *Developing Researcher Independence Through the Hidden Curriculum*,
https://doi.org/10.1007/978-3-031-42875-3_3

scholars' potential development, this chapter mapped their learning opportunities and challenges through the lens of the hidden curriculum, drawing on the Vitae Researcher Development Framework (RDF), a research-driven tool for researchers to reflect on their ability development (Vitae, n.d.; Bray & Boon, 2011). The framework is useful for gaining a systematic overview and common language of researcher development, which is otherwise 'complex and potential(ly) chaotic' (Bray & Boon, 2011, p. 113).

Learning Experience of Doctoral Scholars in Japan

Japanese doctoral education commonly includes completing a dissertation, seminars/coursework, and a viva voce. Candidates must be enrolled in their programme for at least three years (the full-time equivalent). Generally, they have mandatory coursework equivalent to 10 credit points (one credit point equals 34–45 study hours). Their theses are usually written in monograph format. However, today, they are expected to publish one to three peer-reviewed journal articles before submitting their theses. The thesis supervision is conducted individually or as a team. Overall, those in the humanities and social sciences work independently, whereas those in STEM fields often conduct their projects in team-based environments. The general language of instruction is Japanese, but English has been increasingly used when needed, especially in the STEM field.

This chapter focuses on two IDSs' collaborative reflections on their doctoral studies in Japan. The meetings were coordinated by the first author, who is a researcher at a university and who delved into the second and third authors' doctoral experiences. Han Jiawen and Zhang Xun hail from China and will soon submit their theses (at the time of writing). They were engaged in their studies in Japanese. Approximately 30% of the enrolled doctoral scholars were international students in 2020, among whom one-third were from China. Chinese students comprised the largest IDS cohort.

Jiawen began to learn Japanese in college. She then realised the influence of her learned first and foreign languages on her thinking and personality, for example, in how she communicated and built interpersonal relationships. Her experience in an undergraduate double-degree programme between China and Japan stimulated her to proceed with her studies in Japan. She started research on bilingual education, especially in the translanguaging literacy development of bilingual students. She was actively involved in campus academic jobs, such as library assistantship and casual teaching. Upon completing her Ph.D. studies, she will return to

China and continue her research. For that, she first wishes to secure a post in a higher education institution. She will return home for family-related reasons (as she is an only child) and the attractive inducement of employment that the government offers to doctoral returnees.

Another student, Xun, started studying in Japan as a research student after completing her master's degree in China. 'Research student' is a non-regular student status that is used by those willing to pursue an advanced degree after previously studying a different major. Many international students have begun studying in Japan with this status. They are admitted to audit courses but usually cannot accumulate credits. With this status, Xun's first mission was to prepare for a doctoral programme entrance examination. The examinations vary among universities. It typically consists of written essays, oral examinations, or both. In her case, she had an oral examination of her master's thesis and doctoral proposal.

To secure a better academic position in the future, she wishes to study at a well-known national university. Mainly using literature written in English, she wishes to take a post in which she uses English in both research and education, along with Japanese or Chinese. However, her ideal career prospects are unlikely to be fulfilled. As a Japanese language major in her undergraduate studies, she suspects that her future position in China will be related to Japanese language teaching, even though her current research topic, the history of Chinese Christian schools, is distinct.

Our Reflection: Informal Learning Opportunities for International Doctoral Scholars

Our first meeting began with the sharing of Jiawen and Xun's formal doctoral learning opportunities, which subsequently turned to the exploration of their informal learning. As this occurred, we were conscious of being more analytical about the boundary between formal and informal learning opportunities. In this chapter, informal settings refer to learning opportunities that are not institutionally designed as degree requirements and electives. They can be inside or outside universities. Drawing on RDF, we took note of opportunities for informal learning (see Table 1). Using an online spreadsheet, we recorded our reflective dialogue regarding their experiences. They include common experiences, regardless of one's country of origin, but some episodes appear unique to IDSs.

Table 1 Major learning opportunities in the hidden curriculum

Domain A: Knowledge and intellectual abilities	• Writing tutoring → critical thinking skills & interdisciplinary knowledge • Conference participation → deepening subject matter knowledge & research techniques • Informally joining formal undergraduate courses → subject matter knowledge in related fields
Domain B: Personal effectiveness	• Conference participation → scholarly networking • Casual teaching & mentor's support → confidence in teaching • Managing a precarious situation → self-management skills • Informally joining undergraduate courses → familiarising oneself with eminent scholars
Domain C: Research governance and organisation	• Library assistant → project management • Unfavourable overlap of dissertation topics → being attentive to project uniqueness & research planning skills
Domain D: Engagement, influence, and impact	• Conference participation → presentation skills • Writing tutoring → writing skills & team working skills

Domain A: Knowledge and Intellectual Abilities

Domain A of the RDF concerns researchers' depth of knowledge, range of intellectual skills, and research techniques. The two doctoral scholars engaged in informal opportunities unique to their circumstances. For example, Jiawen has served as a peer-writing tutor for several years at her university. As a student tutor, she helps students encountering writing challenges by identifying their problems and finding solutions independently. She feels that she has developed her critical thinking skills and gained novel interdisciplinary knowledge by reading others' writings. This work allowed her to observe students' in-depth thinking processes through collaborative and problem-solving advice.

Conference attendance, while not necessarily obligatory for achieving their degree, was also helpful. Jiawen received both positive comments and criticism from the audience when presenting her research. These opportunities made her aware of vital academic skills, including presentation skills (cf., Domain D) and basic research techniques. Without joining conferences, she believes that her awareness of these skills would not have been enhanced. Furthermore, the conferences helped her build scholarly networks (cf., Domain B) and gain new ideas. As she crossed institutional boundaries at the conferences, she met experts in related fields and learned new trends and future research directions.

Interestingly, Xun informally used formal undergraduate courses in an 'innovative' manner (Elliot et al., 2020, p. 36). She believes that doctoral scholars, especially in the humanities, should acquire broad knowledge of relevant subject matter. Since her doctoral topic differs from those she studied in her prior degrees, she informally sat in on several undergraduate courses. She recollected the enlightenment of analysing some local cases using sociological theories that professors talked about in the courses. Additionally, because her current subject is educational history, she was interested in how undergraduate students study in the courses as an observation of Japanese education culture. The two IDSs sought opportunities for knowledge and intellectual development in proactive and innovative manners.

Domain B: Personal Effectiveness

Domain B, personal effectiveness, includes the personal qualities needed to effectively conduct their research work, such as self-confidence, self-management, and personal network. Jiawen developed scholarly confidence during her service as a casual teacher of academic writing in a Japanese language school. Initially, she was afraid of making mistakes and was unsure about her teaching abilities. However, she learned from senior teachers' feedback, designed teaching plans with them, and eventually overcame her anxiety. She now enjoys preparing for teaching, exerting creativity, and helping the students develop their writing skills.

Xun developed her awareness of self-management while grappling with her precarious situation. As a nationally funded student, she must be admitted to a doctoral programme within two years; otherwise, she will lose her scholarship status. The fear of failure put her under considerable stress, but the situation also enhanced her awareness to achieve her goals. Moreover, for Xun, having a new supervisor meant that she had to pursue a somewhat different topic from what she did in her master's studies. Whenever she had to address new subject matters, she found it challenging to produce output comparable to topics she was familiar with; the required time and effort seemed twice or thrice the original effort. She knew, however, that it was her task to demonstrate her diligence and proactiveness in her research work to her supervisor. In addition, attending undergraduate coursework helped her develop her knowledge base as a researcher, know eminent Japanese scholars in related fields, and broaden her personal network. The two IDSs' experiences of personal effectiveness development differed: Jiawen received scaffolding, whereas Xun developed through her independent experiences. However, both overcame the anxiety embedded in the hidden curriculum.

Domain C: Research Governance and Organisation

Researchers are expected to have the skills to properly manage their projects independently and/or collaboratively. As a library assistant, Jiawen was involved in organising writing seminars and library tours for undergraduate students. Experience in managing these projects was valuable in project planning and the delivery of scholarly duties. She also realised that creativity was key to managing the constant variety of events and generating positive outcomes. She shared ideas for the events with the library staff and other assistants, including those that were valuable to first-year students, the difficulties faced by international students, and ways to effectively solve their problems.

There are other doctoral peers in Xun's lab. At the beginning of her project, she noticed that another student was using the same research materials. She feared that this situation might end up in unhealthy competition and even unethical research conduct. To differentiate her project from that of others, she carefully talked about her plan with her professor before embarking on her project. She now believes that she should communicate well with other lab members to avoid overlapping and make a unique contribution to her field of interest. Both Jiawen and Xun acknowledge their development in interpersonal interactions.

Domain D: Engagement, Influence, and Impact

Domain D comprised skills for impacting and working with others. As Jiawen worked as a writing tutor, she had the chance to hone her writing skills, such as paragraph writing. She also developed collaborative skills, with various opportunities. She worked with a group of professors and students to design academic writing materials where they made good use of their experience and expertise to effectively produce intended outcomes. Working with members from different backgrounds, she learned how to exchange opinions with others. Being familiarised with the collaborative tools of information and communications technology, including Dropbox and Google Meet, ensured effective communication inside and outside of physical workplaces. Jiawen's experience as a library assistant allowed her to learn how to negotiate effectively and share her work with others. She became cognisant of the significance of interpersonal communication in working as a student library staff member to proactively employ individuals' abilities and push new projects forward.

Xun recollected that she paid little attention to making any impact using her expertise. This was because she started studying a new subject in her doctoral studies and was extremely concerned about acquiring new knowledge and immersing herself in the field. She also reflects on her short history of studying in Japan. Her personal network has not developed as much as that of other international students who had been residing in Japan for a longer period. COVID-19 restrictions also prevented her from developing the network as she desired. She stated that if the COVID-19 pandemic had not occurred, she would have attended more conferences, worked more with other peers, and helped organise events led by her supervisor. Xun's relative shortage of opportunities was apparent due to societal constraints in response to the pandemic. Jiawen, in contrast, sustained her efforts by relying on opportunities she had seized before the pandemic.

Reflecting on the Insights from Our Dialogues

Some experiences unique to budding scholars such as IDSs surfaced in our dialogues. For example, Xun observed that when one starts studying abroad and thereby changes academic supervisors, one may have to start over with a new topic, which would demand building new knowledge and interpersonal network from scratch. Since IDSs have few scholarly achievements, their research plans appear susceptible to their supervisors' suggestions. Making an impact in a new country and using a foreign language are also among the biggest challenges. These circumstances made her feel like a 'student' rather than an 'expert'. She believed that the COVID-19 lockdown prevented the hidden curriculum from developing an interpersonal network, which may have mitigated the challenges she faced. Furthermore, some of Jiawen's international friends felt a dearth of authentic teaching opportunities for career development. Jiawen's advisor encouraged her to teach international students and participate in peer writing and tutoring. Although she struggled with teaching Japanese writing as a non-native, less experienced teacher, the informal learning opportunity allowed her to deepen her understanding of her subject matter. Fortunately, she had a supportive mentor for her teaching skill development.

As Elliot et al. (2020) recommended, Xun realised the importance of proactive exploration of informal learning opportunities. While she appreciates university support for international students, she is concerned about international students' passive approach in availing support and opportunities exclusively for international students, as they seem to rely heavily on being fed information without actively seeking broader opportunities for contingent learning. Jiawen also noted that people around her seemed to believe research is solitary work, and she worried they were missing out on the rich learning opportunities in coworking and informal gatherings. In her view, international students seldom participate in informal career development and academic writing seminars (e.g. Elliot et al., 2020). She also believes that support for international students mostly focuses on administrative support, and that local students receive more comprehensive information about support and on-campus opportunities. Undeniably, support should meet students' requirements, but the binary nature of student cohorts appears to create unnecessary boundaries for informal learning opportunities between international and local students. Jiawen felt that some academic opportunities on campus were also useful for international students, but the university administration seemed to share some information exclusively among local students.

Surprisingly, although the two IDSs were of the same generation and from the same country, their informal experiences largely differed. The systematic exploration of individual backgrounds is beyond the scope, but their history of overseas studies (Xun: only doctoral studies vs Jiawen: master's and doctoral studies) appears to be key in the divergence of their experiences. Specifically, Jiawen picked up more opportunities during her longer study history in Japan. Even during the COVID-19 crisis, Jiawen continued her part-time job, whereas Xun encountered the pandemic immediately after her enrolment in her doctoral degree, and many academic events and opportunities remained suspended. Xun had few opportunities to engage in the rich repertoire of the hidden curriculum even though she was enrolled in a much larger university.

Our dialogues also attempted to examine the significance of informal learning for doctoral scholars' career prospects. (See other chapters where there is greater focus on the value of career advancement, e.g. Skakni and Inouye, Hancock.) However, they consistently related their uncertainty in future career possibilities. Until recently, Xun was unaware of the policy dynamics and career pathways in the tertiary and secondary education professions. Jiawen was concerned about whether what she learned in Japan, such as Japanese work ethics and norms, had comparable values in China. Elliot et al. (2020) described the challenges faced by IDSs in transcending two different 'ecological systems', referring to their host and original environments. Soon-to-be returnees to their country of origin were seldom concerned about their prospective situations in their original ecological system. They embody the co-existence of two past and current ecological systems but may not easily envision the third 'future' ecological systems in which they may work.

Studies on IDSs have often shared their challenges and development opportunities in the cultural and linguistic domains (e.g. Elliot et al., 2020). This chapter sheds light on IDSs harnessing the hidden curriculum, which may have contributed to the development of researchers' abilities. Through close interpersonal relationships, the two IDSs acknowledged the role of interdependence and engaged in the rich development opportunities available in the hidden curriculum towards becoming independent researchers. However, like Xun, who was somewhat solitary due to the pandemic, she designed a hidden curriculum on her own. Challenges in envisioning future career trajectories, however, seemed to discourage their proactive decision-making in managing their development opportunities embedded in the hidden curriculum.

Creative exploration in the hidden curriculum may initially create anxiety, but the IDSs' accounts highlight their resilience in overcoming them. We only focused on the two doctoral scholars who are fluent in Japanese language and did not intend to depict all the variations in IDSs' experiences in Japan. In this case, the 'disparity between the original and the new ecological systems' (Elliot et al., 2020, p. 37) may not be that substantial for the two. For example, linguistic challenges may diminish IDSs' engagement in local peer communities (Sakurai et al., 2012). Doctoral scholars in Japan who do not use Japanese as a daily functional language warrant future attention, as they are increasing in number. It would also be valuable to know how they can harness, engage more strategically in learning opportunities and exercise interdependence with those around them given that this is core to their development and doctoral success in Japan.

References

Bray, R., & Boon, S. (2011). Towards a framework for research career development: An evaluation of the UK's Vitae Researcher Development Framework. *International Journal for Researcher Development, 2*(2), 99–116. https://doi.org/10.1108/17597511111212709

Elliot, D. L., Bengtsen, S. S., Guccione, K., & Kobayashi, S. (2020). *The hidden curriculum in doctoral education*. Springer.

Sakurai, Y., Pyhältö, K., & Lindblom-Ylänne, S. (2012). Factors affecting international doctoral students' academic engagement, satisfaction with their studies, and dropping out. *International Journal for Researcher Development, 3*(2), 99–117. https://doi.org/10.1108/17597511311316964

Vitae. (n.d.). *About the Vitae Researcher Development Framework*. Retrieved February 11, 2023, from https://www.vitae.ac.uk/researchers-professional-development/about-the-vitae-researcher-development-framework

Peer Mentoring: A Potential Route to Researcher Independence

Dangeni, *Rui He*, *and Natthaphon Tripornchaisak*

Introduction

Despite the traditional professional research training for doctoral scholars offered by most universities, learning and mentoring opportunities among peer doctoral scholars are less examined, but can serve as crucial elements of doctoral scholars' learning and development processes. Reflecting on the authors' personal and professional hidden learning opportunities from our experiences of working collaboratively as Graduate Teaching Assistants (GTAs) during our international PhD journeys at a UK university, this

Dangeni (✉)
Newcastle University, Newcastle, UK
e-mail: dangeni@newcastle.ac.uk

R. He
University of Manchester, Manchester, UK
e-mail: rui.elin.he@manchester.ac.uk

N. Tripornchaisak
University of Glasgow, Glasgow, UK
e-mail: natthaphon.tripornchaisak@glasgow.ac.uk

D. L. Elliot et al. (eds.), *Developing Researcher Independence Through the Hidden Curriculum*,
https://doi.org/10.1007/978-3-031-42875-3_4

chapter will focus on our development experiences through the lens of peer mentoring. We will discuss how we took the agency while being given flexible teaching instructions, to construct this crucial peer mentoring collaboration, particularly during the pandemic, and then elaborate on what and how we learned and benefited from engaging in this peer mentoring activity, with a focus on post-PhD careers. Ultimately, enhanced agency and engagement in this collaborative peer mentoring process have enabled the successful harnessing of the 'hidden curriculum' leading to the development of both researcher independence and career (see also Aarnikoivu's chapter discussing advancing one's career).

This chapter is based on our reflections when we were working as GTAs for a semester-long research methods course 'Introduction to Educational and Social Research' (IESR) (Dangeni et al., 2021). This course is an introduction to the fundamentals of research philosophy, methods, and practice. In the academic year 2020/2021, it hosted 571 master's-level students. We were each at a different stage of the PhD journey, which also meant that we had different study goals we needed to achieve. For instance, Natthaphon was collecting data for his PhD project, Dangeni was working on her Discussion chapter, and Rui had just passed her viva. Nevertheless, we were able to provide reflective insights from different stages of the PhD journey.

In this chapter, we first discuss peer mentoring to share the value of crucial 'in-practice' collaborative learning opportunities embedded in a GTA group and to voice our experiences. We then exemplify how we harnessed hidden collaborative learning opportunities, before commenting on our deeper understanding of the fruitful gains from this opportune peer mentoring process. Finally, we offer our thoughts on the ways such an experience has contributed to our ongoing development in becoming more independent as researchers.

What Is 'Peer Mentoring'?

Being a doctoral scholar brings diverse and multifaceted opportunities and challenges as one enters a unique academic research culture that requires a new level of learning and knowledge production. In contexts such as the UK, the core of the doctoral learning process tends to be embodied by the regular interactions between the doctoral scholar and supervisor(s) (Parker, 2009). However, given the complexity of doctoral scholars' learning and development, growing attention has been paid to more open and flexible opportunities and approaches to doctoral education, e.g., hidden curriculum (Elliot et al., 2016), community-building (Cai et al., 2019). A key concept of this chapter, peer mentoring not only describes a relationship where more experienced doctoral scholars support the less experienced with advice and knowledge (Collier, 2017; Colvin & Ashman, 2010), but it is also about everyone in the group supporting each other and reflecting on researcher and career development. Typically, peer mentoring relationships are characterised by 'regular/consistent interactions' between individuals (Lorenzetti et al., 2019) and have been highlighted as a way of enabling the provision of psychological support and diverse learning opportunities (Webb et al., 2009). Examples of peer mentoring generally include a more experienced student helping a less experienced student improve academic performance in the university (Colvin & Ashman, 2010) or ethnic student groups as an effective source of support for international students (Colvin & Jaffar, 2007). Yet, there is limited understanding regarding peer mentoring among doctoral scholars who are engaged in teaching. Arguably, such areas warranting attention include: ways in which doctoral scholars initially join a peer mentoring group, seek to develop and reflect upon their teaching practices and experiences, and ultimately experience researcher development. We further argue that doctoral scholars develop their researcher independence through engaging with peer mentoring. For us, facilitating and supporting this research methods course as a GTA team was an unanticipated but golden opportunity to enhance our academic identities by learning from observing each other and reflecting together on our experiences. The following section will contextualise peer mentoring through our engagement with teaching. Through our reflections, we further elaborate on the developmental trajectories contributing to our researcher independence.

Peer Mentoring as a Route to Researcher Independence

GTA Experience as a Secondary Source of Learning

Before the semester started, the course convenor of IESR suggested potential ways in which our GTA team could support students' learning, including providing weekly office hours (i.e., a weekly online drop-in space for students). At the same time, we, as a team, determined how many hours each of us could contribute. Being at different stages of our PhD, we designed and planned a working pattern to maximise our efficiency as a team, e.g., a merged table to assign these office hours for the entire semester. We carefully discussed and prepared a schedule that would not overlap with any of the students' other classes or courses, with a view to encouraging greater student participation. By tailor-making the schedule to suit student timelines and availability, we, as doctoral scholars, learned to make independent decisions within the guided instructions from the course convenor.

During Zoom meetings, we flexibly allocated our responsibilities according to the number of attendees. If there was only one student, all of us would remain in the main room. If more students entered the room, each of us would go to separate breakout rooms. The team would try to answer students' questions while directing the students to their tutors or relevant resources. Sometimes, we received hardly any questions, but at other times, we had a rather active group, especially towards the end of their assignment deadlines, as expected, as students prepared for submission. Another task was to help answer discussion board questions via Moodle. There would be one discussion board for the whole course, and we needed to check it regularly and answer questions. In the indicative task allocation we created, we had one member in charge of the discussion forum while the rest of us would lead the drop-in sessions each week. This meant that even amongst us, we started to take turns exerting our independence while being supported by one another—perhaps, the starting point for a very informal peer mentoring scheme.

Negotiating plans and timetables to achieve our tasks by ourselves was the beginning, instilling in us a real sense of researcher independence by solving most small problems (i.e., operational issues) independently. As an example, many master's students in this course found 'research paradigm' a very challenging term to understand. The three of us then further reflected on, discussed, and shared our understanding of it and some useful resources we used for our PhD learning. On reflection, teaching others spontaneously contributed and enhanced a sense of researcher independence among the three of us. In handling the students' questions in the main and breakout rooms, we also informally mentored each other—further applying these lessons in our respective PhD studies. To illustrate, Natthaphon was able to emulate his senior colleagues' approaches to completing their PhDs. He also learned from the students' questions and how other GTAs handled them, such as questions about data analysis, findings, and discussion. By listening to various applications of creative research methods, e.g., vignettes, visual methods, and diary studies, Dangeni was able to reflect on her chosen research methods, and, in turn, refine her methodology and discussion chapters, i.e., epistemology and ontology. Altogether, discussing and sharing our understandings and experiences proved beneficial not only in supporting master's students but also in supporting our own doctoral research.

Working in Academia as International GTAs

As international doctoral scholars, we were somehow expected to understand more about the 'real world' of teaching and learning provision in the UK Higher Education (HE) context, even before joining this GTA group. Natthaphon already had five years of undergraduate teaching experience in Thailand, and so his initial expectations regarding GTA work were to prepare him further, perhaps for postgraduate teaching responsibilities since these responsibilities would be expected of him after his PhD. Dangeni had previously worked as a GTA, but mainly supporting students' group work and online discussions. She was equally keen to explore the different roles and learning opportunities that a GTA role could provide to engage students given her research interests, i.e., student engagement. For Rui, she aspired to gain more practical experience to support her job-seeking at the end of her PhD. Despite varying personal objectives for what the GTA role could offer us, the shared opportunity to learn how to run a module and initiate a GTA team benefited us equally in terms of broadening our knowledge of working in academia—a destination aspired to by a number of doctoral scholars. Being briefed by the course convenor on the course details, GTA roles, and tasks in general allowed us to learn the basics of setting up a module, e.g., course timeline, handbook, Intended Learning Outcomes (ILOs), launching online learning platforms, and teamwork (among course leader, admin staff, GTAs, and more), which is valuable and meaningful for our career progression, especially for those who want to teach in HE in the future.

Working collaboratively, we took the lead in designing how we wanted to run the online office hours and decided how many hours each of us could contribute. This is considered a vital step and valuable exercise for us as we need to take ownership of our teaching and its management. The planning not only gave us a good foundation for our future teaching with clearly defined roles and responsibilities. (This is a concrete example of what Skakni and Inouye argue to be crucial in planning for post-PhD career aspirations.) Equally, we found that being involved in the planning enabled autonomy and the skills to manage our teaching as doctoral scholars and future academics. We did not expect such an opportunity, especially since some of our previous GTA experiences had only involved some administrative work, with very little systematic design and planning. It was a bit daunting at first, but soon became exciting because of the sense of ownership it gave us. Being at different stages of our PhDs crucially informed how we designed and planned a working pattern that maximised our efficiency as a team, e.g., via creating a merged table with our available hours for the entire semester. One member was responsible for the online discussion forum used to discuss, clarify, ask questions, share ideas, and help students out as and when needed. The rest of us ran the drop-in sessions, where students met with us, the course GTAs, to ask questions about the course and course content, for general study and reading advice, to ask questions about where to find things on our course Moodle page, or to check-in for any other course-related questions. This can be considered an example of optimising our time and task management skills through working in a team, which is useful in developing researcher independence and is an equally essential skill for our career in the long run.

We practised many teaching approaches we had observed, experienced, and researched as effective, engaging, and supportive for students throughout the semester, e.g., feedback and feedforward. We also gained a better understanding of some 'career notions' that were new to us, e.g., drawing boundaries. We found that the ease with which we allowed students to ask questions sometimes led to opportunities for students seeing the GTAs as the final validation for their research topics/dissertations. For example, some of us kept receiving individual emails from a particular student. We discussed this within the team and also sought advice from the course leader. As a result, we decided that we would not make the final decisions but would only provide suggestions and signpost students to useful resources. Additionally, while being a GTA gave us a realistic insight into real-world teaching, we also reflected on the concept of 'independent learning' as doctoral scholars. Independent learning does not mean that you are always learning and researching alone. In contrast, you can always find a way to seek and access support and resources in order to engage with peers and communities to explore the process with you.

Peer Support During Lockdown

Our experience of peer mentoring took place during the UK's third Covid-19 pandemic lockdown, i.e., January–July 2021. Working collaboratively as part of a GTA team, as described comprehensively earlier, also paved the way for our experience of mutual psychological support via peer mentoring. (This is in line with Rainford's chapter discussing 'support villages'.) Surprisingly, but gratefully, our peer mentoring was not limited to academic learning and course-related topics as part of our GTA remit. All three of us found it somewhat difficult to continue writing our thesis or other work efficiently, or to maintain a healthy work-life balance as our lives were constrained by living in a 'small box' (i.e., our houses or studios in student accommodation). Spontaneously, we made the most of our time and took the opportunity to regularly check in on each other's lives and wellbeing. For example, while waiting for our students to join the online Zoom session, we would chat about things happening around us; share our thoughts, feelings, or academic or life challenges; encourage and cheer each other up; and offer practical suggestions for various challenges we were facing. By supporting each other during this difficult time, we all felt greatly supported, both academically and personally. For Rui and Dangeni, in particular, who were living alone in studios in student accommodation, being able to 'see' familiar faces with encouraging smiles greatly helped alleviate our isolation and anxiety.

By the time this chapter is being written, Rui and Dangeni have moved on to academic careers and Natthaphon has also become the 'grandfather' of the School's GTA team (i.e., a senior member who would, for example, explain the overall expectations of being a GTA to newcomers). However, we are all surprised at how much we benefited from this opportune peer mentoring experience during our GTA work, and how interesting it was for the three of us to continue it in different ways. Natthaphon, who just learnt his GTA ropes two years ago from Dangeni and Rui, is now mentoring his junior PGR peers and supporting them in the same way that he was supported. Dangeni now works as a practitioner in HE, supporting doctoral scholars with teaching responsibilities. She is becoming very confident in setting up modules from scratch and designing different ways to support doctoral scholars' learning, engagement, and community-building. Rui now works as an academic and supervisor for both master's and PhD scholars. She also tries to create supportive communities and promote these 'hidden curriculum' opportunities for her students, encouraging them to peer-mentor and support each other.

More importantly, the three of us have developed a strong camaraderie that continues. We still support each other and act as peer mentors, even though two members of the team have moved to different cities. Completing (or about to complete) the PhD journey can be exciting but also daunting as we leave our supervisor's nest and become completely independent. However, this peer mentoring experience from our GTA work and the ongoing peer support reassure us—that we are not alone in the process of becoming more independent scholars. Arguably, interdependence, as demonstrated throughout this chapter, is key.

Takeaway Messages from Peer Mentoring

We hope to share our reflections on our peer mentoring experiences with other doctoral scholars (especially international doctoral scholars), supervisors, and practitioners. We believe that these reflections have meaningful implications for anyone involved in doctoral education. For doctoral scholars, the multifaceted GTA experiences, e.g., familiarising oneself with the curriculum, teaching and communicating with students, as well as peer mentoring, have greatly facilitated our researcher development from multiple perspectives. Our reflections provided solid and vivid evidence of the importance of peer mentoring experiences in the doctoral journey, such as balancing teaching and researcher identity (Collins et al., 2021) or deeper exploration of the subject area (Muzaka, 2009). As we demonstrated, such experiences can also have a significant impact on becoming independent researchers with enhanced problem-solving skills in academic contexts (Lorenzetti et al., 2019) and a safe space for mutual growth. For doctoral scholars who are interested in exploring GTA roles and any other peer mentoring opportunities, we encourage you to give it a go! Initiating a peer mentoring community, whether small or large, is likely to make a difference as you try to make the most of engagement and development in your doctoral journey. First, here are the essentials to bear in mind:

1. Be aware of the opportunities available to you by ensuring that you have navigated the various opportunities in your department/school/university.

2. In creating a peer mentoring community based on any context (e.g., teaching), you need to be prepared not only to receive but also to offer reciprocal support.
3. Peer mentoring prompts reflection on your activities (e.g., teaching) but also on researcher development—individually and collectively.

Supervisors and key stakeholders involved in supporting researcher development (e.g., directors of doctoral scholars, researcher developers) should consider encouraging and supporting doctoral researchers to engage in peer mentoring practices or other forms of community building alongside teaching and research. These activities can arguably help equip doctoral scholars with the necessary skills for their future careers. Opportunities from departmental and institutional communities for and with doctoral scholars, and communities of shared interests and needs (Cai et al., 2019), are all potentially meaningful contributors to the development of researcher independence. Given the rewarding experiences and reflections awaiting doctoral scholars, such as academic development, psychological support, and career suggestions (Lorenzetti et al., 2019), such engagement can powerfully transform the often isolated and challenging journey into a reflexive and developmental one. Researcher independence does not come automatically. One example of hidden curricular learning through peer mentoring, as featured in this chapter, has contributed greatly to our own development and researcher independence journey via proactive planning, learning, and continuously reinforcing each other's learning. Based on our collaborative GTA experience, we contend that this is a meaningful ongoing process for achieving and enhancing researcher independence that is core to our career development.

Declarations of Conflicts of Interest No potential conflict of interest was reported by the authors.

Acknowledgements The authors would like to thank Dr Kara Makara for the opportunity and guidance on GTA work. Thanks are also owed to the editor, Dr Dely Elliot, for the helpful feedback on an earlier draft of this chapter.

References

Cai, L., Dangeni, D., Elliot, D. L., He, R., Liu, J., Makara, K. A., Pacheco, E.-M., Shih, H.-Y., Wang, W., & Zhang, J. (2019). A conceptual enquiry into communities of practice as praxis in international doctoral education. *Journal of Praxis in Higher Education, 1*(1), 11–36. https://doi.org/10.47989/kpdc74

Collins, J., Brown, N., & Leigh, J. (2021). Making sense of cultural bumps: Supporting international graduate teaching assistants with their teaching. *Innovations in Education and Teaching International*, 1–11. https://doi.org/10.1080/14703297.2021.1919175

Collier, P. (2017). Why peer mentoring is an effective approach for promoting college student success. *Metropolitan Universities, 28*(3), 9–19. https://doi.org/10.18060/21539

Colvin, C., & Jaffar, F. (2007, July). Enhancing the international student experience: The importance of international student groups and peer support at Edith Cowan University. *In 10th Pacific Rim First Year in Higher Education Conference.* https://www.researchgate.net/profile/Cassandra-Colvin/publication/237378667_Enhancing_the_International_Student_Experience_The_importance_of_International_Student_Groups_and_peer_support_at_Edith_Cowan_University/links/53f55c540cf2fceacc6f43da/Enhancing-the-International-Student-Experience-The-importance-of-International-Student-Groups-and-peer-support-at-Edith-Cowan-University.pdf

Colvin, J. W., & Ashman, M. (2010). Roles, risks, and benefits of peer mentoring relationships in higher education. *Mentoring & Tutoring: Partnership in Learning, 18*(2), 121–134. https://doi.org/10.1080/13611261003678879

Dangeni,He,R.,&Tripornchaisak,N.(2021,August16).Peermentoring:Becoming independent from teaching collaboratively. *The Hidden Curriculum in Doctoral Education.* https://drhiddencurriculum.wordpress.com/2021/08/16/peer mentoring-becoming-independent-from-teaching-collaboratively/

Elliot, D. L., Baumfield, V., Reid, K., & Makara, K. A. (2016). Hidden treasure: Successful international doctoral students who found and harnessed the hidden curriculum. *Oxford Review of Education, 42*(6), 733–748. https://10.1080/03054985.2016.1229664

Lorenzetti, D. L., Shipton, L., Nowell, L., Jacobsen, M., Lorenzetti, L., Clancy, T., & Paolucci, E. O. (2019). A systematic review of graduate student peer mentorship in academia. *Mentoring & Tutoring: Partnership in Learning, 27*(5), 549–576. https://doi.org/10.1080/13611267.2019.1686694

Muzaka, V. (2009). The niche of Graduate Teaching Assistants (GTAs): Perceptions and reflections. *Teaching in Higher Education, 14*(1), 1–12. https://doi.org/10.1080/13562510802602400

Parker, R. (2009). A learning community approach to doctoral education in the social sciences. *Teaching in Higher Education, 14*(1), 43–54. https://doi.org/10.1080/13562510802602533

Webb, A. K., Wangmo, T., Ewen, H. H., Teaster, P. B., & Hatch, L. R. (2009). Peer and faculty mentoring for students pursuing a PhD in gerontology. *Educational Gerontology, 35*(12), 1089–1106. https://doi.org/10.1080/03601270902917869

'Settle Down' or 'Return Overseas'? A Reflexive Narrative of an International Chinese Doctoral Scholar's (Re)adjustment Experiences in China

Kun Dai

Introduction

Among international students, the Chinese group is one of the largest cohorts in many countries, such as the United States, the United Kingdom, and Australia. Although Chinese international returnees' experiences have been initially explored, as Jiang et al. (2020) suggested, limited studies have explicitly examined Chinese doctoral scholars' employability and illustrated the trajectory of job acquisition as early-career researchers who may encounter a very competitive situation in both domestic and international job markets. (Similar concerns have been raised in other chapters, e.g. Skakni and Inouye, Hancock.) Against the background introduced

K. Dai (✉)
Department of Educational Administration and Policy, Faculty of Education, The Chinese University of Hong Kong, Hong Kong, China
e-mail: kundai@cuhk.edu.hk

D. L. Elliot et al. (eds.), *Developing Researcher Independence Through the Hidden Curriculum*,
https://doi.org/10.1007/978-3-031-42875-3_5

above, I have attempted to analyse reflexively and illustrate my (re)adjustment and job acquisition experiences as an international doctoral scholar from an Australian university. Moreover, as Elliot et al. (2016) suggested, doctoral scholars not only learn scientific knowledge but also immerse in a process of acquiring 'hidden treasure', which may significantly influence learning experiences and academic development. Thus, 'hidden curriculum' that I experienced during my transitioning journey will be illustrated.

This reflexive narrative provides nuanced illustration of my job acquisition across different higher education (HE) job markets. This approach is a way of autobiographical storytelling. Specifically, I used Bourdieu's concepts, field, habitus, and capital to interpret my experiences. The exploration of one's individual experience in a reflexive approach is a way of self-assessment and self-reflection (Bourdieu & Wacquant, 1992).

The process of job acquisition may position me in different fields. The habitus and capital I gain may also influence my job acquisition. Bourdieu's concepts could provide a systematic toolkit to explore the complex job acquisition experience in and across different job markets (fields) that have different logics of practice, which refers to norms and rules that people may follow in their everyday work and life in any field. The concept of field usually refers to a social space. For habitus, Bourdieu believes that it is the way that everyone has thought of as the embodiment and sedimentation of past experiences in different fields. Furthermore, according to Bourdieu (1986), there are various types of capital that could influence individuals' behaviours, social mobility, and other everyday activities, typically including, cultural, economic, and social capital, which could be converted with each other. These concepts help me understand the way I adapted to navigate different rules and norms in cross-field movement and the influence of norms and rules on my development as an independent researcher in either hidden or apparent aspects.

The data are based on my research notes and diaries from 2018 to 2020. I wrote my experiences frequently as diaries in my personal online blog. The original purpose was to record my learning and research journey for my future reminiscence. To clarify the storyline, I recursively analysed this data both deductively, drawing on Bourdieu's thinking tools and inductively, generating themes from the data, to identify the specific changes made as a Chinese doctoral scholar. This narrative is mainly based on my individual experiences rather than generalisable to other cases.

Hidden Rules in the Job-Hunting Journey

After completing my doctoral study in late 2018, I prepared to start my new journey. Obtaining a doctoral degree from one of the top universities in Australia and publishing several research articles enhanced my confidence to find an academic job. However, in the process of job hunting, I noticed that the reality is not optimistic.

> I applied to several positions in Australia; however, I always got rejections or no responses. When I communicated with other peers and academics, I felt that education and practices in Australia seemed to mainly focus on early childhood and school levels. Most job positions in Australia focus on school education, which is not my field.

Although I obtained my doctoral degree in Australia and studied with well-known scholars, it seems that my 'so-called' cultural/social capital in the Australian HE fields could not support me to find a suitable position. I did not give up and I tried to learn more about the context and usually went to seminars related to doctoral scholars' job search.

> In a seminar about doctoral scholars' career development, the lecturer mentioned that beyond 50% of jobs in Australian academic market were based on social network. I felt that it is so difficult to find something here as I did not have such a strong network. I also realised that finding a job in market is much harder than learning in university. During the doctoral study, my goal was to graduate and get the degree. I thought that will be the end. However, everything just begins; getting the degree is the starting point rather than the end.

After having more understanding about these hidden rules in the Australian HE field, I noticed that the 'cultural capital' (e.g., a doctoral degree) I obtained may not be sufficient to help me to develop a career. Then, I started to search opportunities in China. Through searching and reviewing many recruiting advertisements, I found that although my backgrounds (e.g., qualification, publication, and age) met several universities' selection criteria, I still needed to have postdoctoral working experiences to apply for positions at reputed universities. Previously, holding an internationally recognised doctoral degree could help doctoral scholars to gain an academic position in China. However, nowadays, the job market has become much more competitive, and the value of foreign degree may also be different in the current era.

Thus, I proposed to apply a research fellow programme offered by a national funding agency. However, I did need to have a host university and advisor to support the application. A senior academic that I met at a conference before from one of the top Chinese universities accepted my request to support the application. Compared to the Australian situation, I noticed that the cultural capital I acquired seemed to be much more recognised in the Chinese HE field, which aims to attract young Chinese doctoral scholars to work in China as a way of improving the internationalisation of HE and research production in the global academic field.

> The programme only allows doctoral scholars from the top 100 foreign universities to apply. However, domestic scholars cannot apply for this fellowship. If I can get it, I can have a better salary package compared to many other postdocs.

Consequently, I was awarded and moved to the university. Although the cultural capital gained from my international learning journey supported me to secure the position that could bring more economic (e.g., high salary package), social (e.g., opportunities to build networks), and symbolic capitals (e.g., university reputation in the Chinese society), I encountered 'misfits' between the habitus that I gained from 10 years study in Australia and the new logic of practices in the Chinese HE field.

Hidden Knowledge to (Re)shape Academic Habitus in the 'Home Field'

Notably, when I moved back to China, I also began the journey of (re) shaping academic habitus in the Chinese HE field. Theoretically, as a student who originally comes from China, I should not have had many issues in adjusting to the Chinese HE field. However, I encountered various challenges and barriers in the initial stage. On the one hand, I was not familiar with many academic policies, norms, and discourses.

> Through communicating with some academics, I noticed that I seemed to have no idea about how to be an academic in China. When they talked about the developmental trajectory in the field, I even did not know what I should do and did not know many particular terms. It seems that I am an

> outsider to the system. I did not have a clear understanding about 'the rules of the game'.

This experience could suggest that I was 'a fish out of water'. The lack of understanding about the logic of practices (both obvious and hidden rules) in the Chinese HE field negatively influenced my confidence even though I got the postdoctoral fellowship at a top Chinese university. Through learning from other colleagues, I smoothly became familiar with some of the rules of the game. In the Chinese HE research field, there are many 'hidden' logics of practices that significantly influence academics' development. The more I knew about the field and logics of practices, the more I felt stressed and struggled. Compared to others who studied in China or had strong social networks with domestic scholars, as a newcomer, I did need to explore the field in a more 'independent' way. Due to a lack of previous connections to significant others in the Chinese HE field, it seems that I did not have strong 'social capital' to support my (re) adjustment to the Chinese system.

On the other hand, I seemed to not have 'academic habitus' that makes me a 'fish in water'. It seems that the habitus that I developed in the Australian HE field mismatch the logic of practices in China. These issues could suggest that when a habitus enters an unfamiliar field, contradictions and uncertainty may emerge as a result of habitus-field misfit (Stahl et al., 2023; Reay, 2004). As I did not have enough research training in China, I did not have a suitable academic habitus in the Chinese HE field. To develop new habitus, I paid more attention to learn knowledge related to the Chinese HE field, for example, academic writing in Chinese. However, the focus on the Chinese field was not enough.

In-betweenness Through Learning Hidden Knowledge

Conducting educational research in the Chinese HE field potentially shaped me as an in-betweener who shifts between different academic fields. As mentioned above, I needed to develop a habitus that enabled a better 'fit' with the Chinese HE field.

> Not only do I need to write research funding proposals in Chinese, but I also need to write some academic articles in Chinese. My advisor told me told that I should publish something in Chinese journals in order to establish

> my domestic reputation; otherwise, I may not be well recognised by others. In China, if I only publish in English, it will be strange. However, publishing in China is also challenging for me. I need to be familiar with features of target journals and pick up writing skills as soon as possible.

Meanwhile, I still need to keep publishing in English to prepare for potential job-hunting pressure in the future. These 'hidden knowledge' in the Chinese HE field pushes me to adapt myself to working in any context. I could take advantage of the cultural capital gained from Australian learning experiences to engage in the international academic field. Meanwhile, I need to foster my Chinese academic habitus to fit into the Chinese system. Thus, I simultaneously conducted my research by following two logics of practice. Those micro shifts in everyday research life potentially position me between different academic fields, which could be seen as a way of developing an in-between, transnational, and cosmopolitan academic habitus: 'For the future, I am still uncertain where I should go: working in China as a doctoral returnee (Hai Gui) or seeking opportunities back to overseas (Gui Hai).' These experiences suggest that a mismatch between habitus and fields exists in the transnational movement, which also creates opportunities for me to strategically maintain and develop different capital. These complex shifts and changes may systematically shape a sense of in-betweenness—developing various senses of identity, agency, and belonging (Dai, 2020, 2022). Meanwhile, while I encountered various challenges and barriers in the journey, the exploration also allowed me to gain hidden knowledge as an international doctoral scholar (Elliot et al., 2016; Elliot et al., 2020). Such an in-betweenness may become a key capital for me to develop myself as an independent researcher familiar with both international and domestic 'rules of the game'.

Discussion and Conclusion

This reflexive narrative provided a nuanced analysis of an international doctoral scholar's trajectory of job-hunting journeys between so-called 'home' (e.g., Australia) and 'host' (e.g., China) contexts. It illustrated a case distinct from many previous studies focused on international scholars' employment and (re)adjustment in either 'home' or 'host' countries. Drawing upon Bourdieu's thinking tools, this narrative could reflect that international doctoral scholars may need to be familiar with different rules of the game in either hidden or obvious aspects to navigate different academic contexts. (The same sentiment is expressed by Sakurai et al.) The author strategically shifted between different rules of the game and hidden curriculum to develop his academic career, which shaped a sense of in-betweenness. This sense could become an essential capital for international doctoral scholars to achieve academic development in the global context. Understanding the hidden curriculum and rules can contribute towards researcher independence in cross-system movement and career development.

Various unexpected issues emerged from 'reality' regarding his navigating journey between different HE job markets. The 'ideal imagination' did not meet the 'real needs' of the Australian academic job market. For more than 10 years, learning in Australia has equipped him with various capitals, including degrees and knowledge about the local context, which are essential for international scholars in Australia (see Pham et al., 2023). However, he seemed not to be familiar with the Australian academic job market's requirements and context. This issue means that he lacked enough understanding of hidden knowledge about the system. These experiences are consistent with previous findings mentioned by Blackmore et al. (2017) who suggested that understanding the labour market's rules is essential for international graduates to find a position in the host country. Moreover, the lack of enough 'social capital' in the Australian HE sub-field also hindered his possible employment. This finding could indicate that 'significant others' and social networks played essential roles in securing academic positions in Australia. In other words, whereas the scholar's sense of agency is crucial in navigating academic contexts either in the host country or at home, it is equally important to recognise the critical role that interdependence often plays in becoming successful in one's career progression.

In contrast, his cultural capital (e.g., an international doctoral degree and publications) was legitimised by the Chinese HE sub-field. In the past decades, China has been actively recruiting international academic returnees to internationalise the Chinese HE field through its economic capital, which is a way of imbricating with the global HE field. Such a logic of practice provides more opportunities for some scholars to develop their academic career. Meanwhile, the significant other (e.g., network with the senior scholar) played an essential role in his success in gaining the national postdoctoral fellowship. These findings reflect that the local networks and individuals' academic performances significantly influenced returnees' employment (Pham & Saito, 2020; Xu, 2021).

However, the author encountered various 'reverse shocks' as a returnee who did not have local education and research experiences. Distinct from Ai (2019) and Xu (2021), the author gained his HE degrees in Australia. Many 'reversed shocks' experienced by those who had previous Chinese learning and research experiences seemed to be 'direct shocks' for him. While he is a 'Chinese' and a doctoral returnee, he appeared to be a 'foreigner' (or 'stranger') in the Chinese HE field. As a result, he needed to learn hidden rules and knowledge to progress his academic career. The (re) adjustment to the Chinese HE context was a tortuous process that progressively shaped his sense of in-betweenness. In this process, he potentially became an in-betweener who may shape an in-between habitus with multiple senses of identity, agency, and belonging. Understanding different hidden and obvious rules in different fields could be essential for some international doctoral scholars to become independent researchers who can move between different academic contexts (see Dai & Hardy, 2023).

These empirical insights could help prospective international doctoral scholars understand the 'realities' in different academic job markets. It is essential for those who aim to find a suitable position in academic fields to develop a rounded capital during their doctoral training. Meanwhile, it is also important to learn informal knowledge (e.g., job market features, policies, and rules) as 'hidden treasure', which significantly influences learning and future development (Elliot et al., 2016; Elliot et al., 2020). Such hidden curriculum could be an important platform for interdependence and subsequently researcher independence development. Universities may also need to provide more professional trainings for doctoral scholars who may need to work in different academic markets.

References

Ai, B. (2019). Pains and gains of working in Chinese universities: An academic returnee's journey. *Higher Education Research & Development, 38*(4), 661–673.

Blackmore, J., Gribble, C., & Rahimi, M. (2017). International education, the formation of capital and graduate employment: Chinese accounting graduates' experiences of the Australian labour market. *Critical Studies in Education, 58*(1), 69–88.

Bourdieu, P. (1986). The forms of capital. In J. G. Richardson (Ed.), *Handbook of theory and research for the sociology of education* (pp. 241–260). Greenwood Press.

Bourdieu, P., & Wacquant, L. J. D. (1992). *An invitation to reflexive sociology*. University of Chicago Press.

Dai, K. (2020). Learning between two systems: A Chinese student's reflexive narrative in a China-Australia articulation programme. *Compare: A Journal of Comparative and International Education, 50*(3), 371–390.

Dai, K. (2022). *Transitioning 'in-between': Chinese Students' Navigating Experiences in Transnational Higher Education Programmes*. Brill.

Dai, K., & Hardy, I. (2023). Contextualizing and hybridizing academic identity formation: An analysis of international returnees and locally trained scholars in China. *Higher Education*. https://doi.org/10.1007/s10734-023-01108-7

Elliot, D. L., Baumfield, V., Reid, K., & Makara, K. A. (2016). Hidden treasure: Successful international doctoral students who found and harnessed the hidden curriculum. *Oxford Review of Education, 42*(6), 733–748.

Elliot, D. L., Bengtsen, S. S. E., Guccione, K., & Kobayashi, S. (2020). *The hidden curriculum in doctoral education*. Palgrave Macmillan.

Jiang, J., Mok, K. H., & Shen, W. (2020). Riding over the national and global disequilibria: International learning and academic career development of Chinese Ph.D. returnees. *Higher Education Policy, 33*(3), 531–554.

Pham, T., & Saito, E. (2020). Career development of returnees: Experienced constraints and navigating strategies of returnees in Vietnam. *Journal of Further and Higher Education, 44*(8), 1052–1064.

Pham, T., Dai, K., & Saito, E. (2023). Forms of agency enacted by international Ph.D. holders in Australia and Ph.D. returnees in China to negotiate employability. *Compare: A Journal of Comparative and International Education*. https://doi.org/10.1080/03057925.2023.2237885

Reay, D. (2004). 'It's all becoming a habitus': Beyond the habitual use of habitus in educational research. *British Journal of Sociology of Education, 25*(4), 431–444.

Stahl, G., Soong, H., Mu, G. M., & Dai, K. (2023). A fish in many waters? Addressing transnational habitus and the reworking of bourdieu in global contexts. *Sociological Research Online*. https://doi.org/10.1177/13607804231180021

Xu, Y. (2021). An autoethnographic narrative of the diasporic experience of a Chinese female Ph. D. returnee's entry into the domestic academic job market during Covid-19: An ecological environment perspective. *Asia Pacific Education Review*, 1–11.

'It Is a Nice Way to End the Week': Journal Club as an Authentic and Safe Learning Space

Kara A. Makara, Dayana Balgabekova, Karen Gordon, Fanzhu Meng, Natthaphon Tripornchaisak, Zhihan Wu, and Dely L. Elliot

The doctoral journey involves the development of significant research skills, one being the ability to critically engage with the literature in one's field. One way that doctoral scholars can better develop this imperative skill is through participation in a Journal Club (JC)—also endorsed in Frick's chapter as a form of environment that strongly fosters creativity development. Often informal and unofficial in nature, these can be considered 'hidden curriculum' within doctoral education through serving as

K. A. Makara (✉) • D. Balgabekova • K. Gordon • F. Meng • N. Tripornchaisak • Z. Wu • D. L. Elliot
University of Glasgow, Glasgow, UK
e-mail: Kara.MakaraFuller@glasgow.ac.uk; Dayana.Balgabekova@glasgow.ac.uk; k.gordon.1@research.gla.ac.uk; f.meng.1@research.gla.ac.uk; n.tripornchaisak.1@research.gla.ac.uk; z.wu.1@research.gla.ac.uk; Dely.Elliot@glasgow.ac.uk

D. L. Elliot et al. (eds.), *Developing Researcher Independence Through the Hidden Curriculum*,
https://doi.org/10.1007/978-3-031-42875-3_6

authentic pedagogical spaces for learning (Elliot et al., 2020). Through collectively discussing and critiquing articles, JCs offer an interdependent space in which scholars arguably learn together about the more implicitly hidden approaches for independent critical engagement with literature that are expected of postgraduate researchers (see relevant concepts in Wisker's chapter, e.g. Fridaying.) While widely used within medical and health fields, JCs are less common and not usually part of the official curriculum within the solo pursuit of doctoral scholarship in the Social Sciences and Education. The limited empirical understanding of how JCs shape doctoral experience warrants an exploration of their potential benefits for these scholars. We, therefore, conducted an autoethnographic analysis of our participation within a JC as a multidisciplinary group of doctoral scholars in Education. Our exploration was guided by the following question: What does being part of a monthly journal club mean for developing researcher independence of a group of doctoral scholars?

Research on Journal Clubs

One approach helping doctoral scholars reach their academic goals more effectively is Journal Clubs (JCs). JCs involve a group of scholars who meet to read and critically discuss articles from the academic literature. JCs offer flexible paths for students to take turns being the leader (i.e. the designated person who facilitates the session) while other participants contribute to the discussion (Swift, 2004). It is a form of interdependent learning whereby peers rely on and mutually benefit from their exchanges with one another. Scholars have found that employing JCs with multidisciplinary participants can lead to increased academic proficiency and prowess in learners (Honey & Baker, 2011). In the doctoral context, the potential benefits of interdependent JC participation on independent learning would be beneficial to understand, as developing critical appraisal and analysis skills are central to completing a doctorate.

While research demonstrates the various benefits of engaging with a JC, some scholars question their usefulness, arguing that JCs are less effective than self-assessment and have limited influence on critical appraisal skills (Alguire, 1998). Sidorov's (1995) survey on JCs within the medical profession across 131 residency programs in America highlighted necessary factors for their effectiveness. First, attendance rates improved for JCs when run independently of the faculty but supported by them, and second, mandatory attendance and benefits such as making food available add to their effectiveness. How these recommendations may benefit PhD scholars' participation in JCs remains unclear.

Swift (2004) argues that the environment plays an important role—beneficial outcomes are reduced if members turn up late or do not engage with the group. Swift also argued for the importance of a responsible leader for overall effectiveness. Within higher education, JCs serve as a complementary component for a variety of purposes, such as enhancing academic writing (Good & McIntyre, 2015) or encouraging a community of practice (Newswander & Borrego, 2009). However, as published studies of JCs have mainly been carried out in clinical and medical fields (e.g. Good & McIntyre, 2015; Harris et al., 2011; Honey & Baker, 2011), the research is limited. More thorough research is warranted to understand the role of JCs for doctoral students' development as researchers.

Reciprocal Determinism as a Lens for Exploring Doctoral Development

Reciprocal Determinism (RD) offers a useful framework for how JCs may support the development of researcher independence. This theory underlines that human functioning results from the mutual and continuous interaction of personal, behavioural, and environmental factors (Bandura, 2001). Personal factors entail an individual's personality traits, beliefs, attitudes, and cognitive processes; behaviour pertains to an individual's actions and learning; and environmental factors encompass the physical and social contexts that influence behaviour. RD underscores the multifaceted nature of human behaviour, which is shaped not only by personality or environment but by their continuous interaction. The theory highlights the pivotal role of personal factors, which dynamically shape individuals' behaviour, with the environment reciprocating these influences (Bandura, 1978, 2001). Comprehending the intricate interrelationships between these factors is crucial to gain a comprehensive understanding of human behaviour.

Within the context of our own study that we carried out on JC, and drawing from this theory, behaviour can be considered as the contributions individuals make to the JC, their conduct during its proceedings, and the resultant changes in their learning. Personal factors encompass individual attributes that affect their participation and personal growth. Furthermore, the environment refers to contextual factors of the JC setting that exert an influence. Doctoral scholars' independent learning and involvement (behaviour) may be influenced by both their attributes (personal factors) and the interdependent group dynamic (environment), and vice versa. For instance, participants reading the article beforehand (behaviour) may learn from their interactive contributions to the leader's questions, fostering a positive learning environment (environment) and motivating the individuals involved (personal). Because RD facilitates bidirectional interactions between doctoral scholars' learning experiences and their personal characteristics situated within the interdependent JC environment, it offers a framework for our exploration of JCs on developing researcher independence.

An Autoethnographic Exploration of Our Own Journal Club Participation

We centred our exploration around our own JC, where six doctoral scholars and two academics met on the last Friday of each month to have a focused discussion on a participant's choice of article. Our JC, one of the activities within our monthly doctoral group meetings, was additional to regular individual supervision meetings (see Elliot & Makara, 2021). Taking turns to lead each discussion was key. An article was chosen by the leader and shared with the group a week before each meeting to encourage advanced reading and reflection. The discussion of each article was primarily guided by a list of questions (see Appendix). Such questions aim to cultivate good thinking and academic judgement, as explained by Peseta et al. in their chapter.

In order to understand how JC shaped developing researcher independence, we explored individual reflections that we wrote while participating in this JC. The autoethnographic approach we took authorised us to draw from our subjective and personal experiences (Bochner & Ellis, 2022) and write 'in a highly personalized style', with a view to enhancing understanding of a specific phenomenon. Autoethnography emerges from ethnographic traditions, a distinct form of research that is 'concerned with the ordinary' and aims to generate an in-depth understanding of a phenomenon by seeking a group's perspective in their 'natural' settings (Harding, 2019, p. 35). Autoethnography's empowering genre is in line with the notion of personal narratives, lived experiences, or reflexive writing (Wall, 2006, pp. 146–149).

Due to the autoethnographic nature of the exploration that we undertook, it did not require ethics approval. However, we carefully considered rigour and ethics through being transparent about the research process (Ellis & Bochner, 2000) and taking collective responsibility for data processing and analysis. Additionally, we used pseudonyms to protect our real identities. Table 1 shows the JC participants.

Over five months of taking reflections, we generated 24 monthly written reflections from seven members. In order to make sense of the reflections, an inductive thematic approach (Braun & Clarke, 2021a; Braun & Clarke, 2021b) was employed using these steps:

- Preliminary analysis of a set of reflections in pairs to identify patterns grounded in the data. This involved carefully reading the reflections line by line, colour-coding the written text in pairs, and cross-checking the codes with the other paired researchers.
- Individual analysis of a set of three to four reflections each.
- A group discussion on the emergent themes through review of codes, trends, and patterns.
- Collective cross-checking of the themes to see whether the themes fall into specific categories for defining the main themes and sub-themes.

Through iterative analysis to refine these themes and ensure collective agreement, we arrived at six sub-themes, grouped into the three components of the Reciprocal Determinism model.

Table 1 Participant profile

Pseudonym	*Gender*	*Role*
Su	Female	1st year PGR
Melissa	Female	2nd year PGR
Emily	Female	2nd year PGR
Irn	Male	Final year PGR
Aibike	Female	Final year PGR
Madelyn	Female	Supervisor
Gabrielle	Female	Supervisor

Multiple Dimensions of Learning Through Journal Club

Our exploration suggests that JC supports multiple dimensions of learning, both (1A) academic learning and (1B) personal learning. Participants in the Journal Club (JC) reported *academic learning* as one key dimension of their experience, referring to learning about different aspects of research, for example, research methods, writing empirical journal articles, and critically reviewing research. Engaging with different studies through JC helped participants gain a deeper understanding of specific methods, such as 'using a creative research method' (Aibike), or more general understanding of research, including 'insights into the core values of delivering good research' (Irn).

> *It allowed us to explore a new component of academic writing (journal requirements in other disciplines) and to explore a different analytical approach (content analysis).* (Madelyn)

Furthermore, participants reported that the JC helped them develop academic writing skills and they were able to learn from the writing style of the articles they read. For example, Su noted, 'I learned that clearly introducing and identifying a theory and constructs/concepts in a paper was critical'. This was echoed by the supervisors as well, with Gabrielle reflecting that the JC discussion led to a 'greater awareness of the standards required for a published paper'.

Engagement in JC also helped participants gain new insights into themselves. *Personal learning* refers to new insights and reflections about oneself in terms of research, choices, beliefs, and attitudes. Participants wrote that the JC provided a space for self-reflection, and discussions at the JC challenged their assumptions and helped them grow as researchers. Both Emily and Irn offered examples of such gradual but explicit improvement.

> *I will need to be able to defend my positions, without letting the emotional connection to things which are part of a lived experience prevent me from doing this professionally and respectfully. It is a challenge, but it is a necessary one and I hope that I continue to improve in this area.* (Emily)

> *...discussing resilience gave me the opportunity to think about my own journey as a PhD researcher who was already in a later year than my peers while also cross-benefitting the development of my ideas on another project on the [PGR] resilience.* (Irn)

Personal Aspects Relevant to the Journal Club Experience

Personal aspects, including (2A) evolving emotions and (2B) sparking a personal link, were also important aspects of developing researcher independence. *Evolving emotions* refers to how JC participation elicited a range of emotions, which changed throughout the process. Participants' emotions were heightened during JC, and they experienced a diverse range of emotions, from stress and worry to excitement and confusion.

> *Leading up to this meeting, I was actually really nervous. ... Stressed! Worried about offending anyone. ... I felt proud of myself after finishing the seminar that I didn't disregard my own beliefs to make the conversation "easier" as this would have left me feeling awful after. So although I found it a very difficult process, it was also rewarding. And being able to have these learning moments in a welcoming space with my peers was calming.* (Emily)

Participants reported that they felt safe expressing their emotions in the JC as there was no fear of judgment. Instead, JC provided a supportive environment that allowed them to express both positive and negative emotions. This was articulated by Su who shared her doubts concerning participation in the journal article discussions:

> *Although it was sort of stressful for me to prepare for and attend an academic discussion because negative thoughts such as 'you're not good enough to do it' or 'you don't have the skills to contribute' occurred, it was beneficial to slow down and examine such thinking process while simultaneously seeking out the positive aspects—the actual situation may not be as bad as I had imagined.* (Su)

Participants also highlighted that the JC *sparked a personal link*, which involved them making connections between the JC experience and their own research. Discussions within the JC felt personally relevant and they were able to apply what they learned to their own research:

> *Although the article is not in my research area, there are still some points to inspire me. For example, the authors used google form to recruit participants online, which gave me a new thought about the recruitment of my research.* (Melissa)

Participants noted that the JC helped them meaningfully reflect upon their own doctoral research and make connections to other research areas. This can helpfully encourage their sense of developing academic identity. One of the supervisors, Gabrielle, was excited to see the doctoral scholars making these connections: 'I was eager to see how members made the connections between the paper and their own research. It's a useful activity for "joining the dots".'

Experiencing the Environment of the Journal Club

Based on our exploration, influences from the environment of the JC included a (3A) supportive pedagogical approach and (3B) positive group dynamics. Participants reported that the JC provided a *supportive pedagogical approach* that was conducive to their learning. The use of guiding questions for leading the discussion was mentioned by several as a useful instructional guide for helping to focus on what matters within the articles.

> *...by deeply engaging with the guiding questions and trying to find answers to them in this article, I have developed a better understanding of the structure/elements of a good academic article.* (Aibike)

Participants also noted that the JC provided a welcoming pedagogical space that allowed for active interaction, perspective sharing, collective learning, and discussion. The concept of space was mentioned several times:

> *This isn't something that would have necessarily been apparent to me without having the space to discuss things in this informal way.* (Emily)

> *It's a safe space to put forward their views, to challenge ideas and concepts from the authors whose work they read.* (Gabrielle)

Positive group dynamics were also a key feature of the JC, referring to when participants found themselves in a friendly and relaxing environment with one another that supported engagement, and, in turn, led to enjoyable and productive working. To experience positive group dynamics, a combination of several factors is seemingly essential. First, almost everyone highlighted the added value of seeing and interacting with one another in person. Second, they commented favourably on having 'a welcoming space' to assist active interaction and discussion—including topics that invite 'opposing opinions'.

Third, the regularity of the JC meetings embedded them into these scholars' routine activities. In both supervisor and doctoral scholar passages, the Friday JC meetings have become something members look forward to—'like a treat' after a long, busy week. Madelyn, a supervisor, noted: 'I like having the journal club on a Friday, because all week long at work I am dealing with a lot ... it is a nice way to end the week'. The sentiment was echoed by the doctoral scholars as well:

> *The seminar itself was very enjoyable. ... We even got snacks which always puts a smile on my face! I also enjoy that the sessions are on a Friday. After a long week of GTA work, intern work, meetings and writing, ending the week seeing familiar and friendly faces feels like a treat.* (Emily)

JC as an Authentic and Safe Learning Space

While JCs are scholarly activities, they are neither considered as 'curriculum proper' nor something in which doctoral scholars typically engage. As in our case, JCs are often not institutionally supported and instead, are informally created leading to an incidental form of learning. Despite being confined within the hidden curriculum, they arguably channel genuine doctoral pedagogies (Elliot et al., 2020), sharing similar principles with 'constellation mentoring' where members act as mentors and mentees and mutually support each other (Li et al., 2018, p. 567). The platform offers space for socialisation—invaluable in doctoral learning—and serves as 'a source of reciprocal learning and enrichment' (Elliot, 2023, p. 114). Within this authentic learning space, not only are multiple dimensions of learning fostered, but a wide range of emotions are generated. Although members reported a wide array of heightened emotions, these were displayed in a 'safe' and 'enjoyable' context. This then enabled members to make connections between JC activities and their own research. Supportive pedagogical approaches including the guide questions and the positive group dynamics reinforced participants' learning experience.

Taking Forward JCs for Developing Researcher Independence

Bringing the findings from our exploration together, we offer a model (see Fig. 1), building upon the framework of Reciprocal Determinism (Bandura, 2001), to represent the mutual interactions found between our learning through the JC, personal aspects relevant to our participation, and the environment of the JC. In this way, our exploration suggests that the developing researcher independence of doctoral scholars can be thought of as interdependence through continuous and mutual interactions with one another within the environment of the JC.

The findings contribute to the literature on the benefits of JCs (Honey & Baker, 2011), expanding this to the context of doctoral study in Education, and suggesting informal JCs offer a valuable and safe interdependent learning space for doctoral scholars' independent research development. The organisation of such 'hidden' learning activities enhances scholars' comprehension of theories, methods, arguments, and academic writings. A further benefit of JCs is the potential for supervisors to encourage their supervisees to take part in scholarly debates and arguments, outwith regular supervision sessions. Supportive instructions such as guide questions can be employed to enrich the pedagogies employed. In contrast to Sidorov (1995), our findings did not suggest a hindrance from

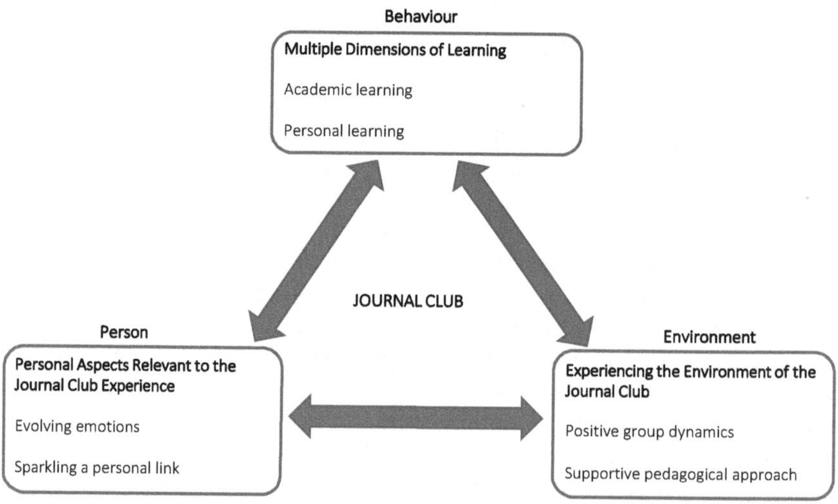

Fig. 1 An adapted model of reciprocal determinism: doctoral scholars' independent learning through interdependent participation in a Journal Club

participation alongside supervisors, yet we did find similar benefits regarding the provision of food. However, this would not preclude doctoral scholars independently designing their own JCs, perhaps where more senior members could take initiative to scaffold the activity for their peers. Finally, institutions can offer support through facilitating appropriate comfortable learning spaces and ensuring JC remains voluntary and one's contributions are not judged.

Taken together, whereas JCs are likely to exist on the periphery of doctoral education, our autoethnographic evidence contends that the doctoral experience emanating from them is instrumental in developing scholars' researcher competence and independence. As a fun and safe space, JCs can offer the opportunity for authentic collaborative learning, including but not limited to the building of doctoral scholars' understanding of how to critically engage with the research literature in one's field.

Declarations of Conflicts of Interest Nothing to declare.

Acknowledgements We wish to thank our family and friends who support us as we navigate through our PhD journeys and careers in academia.

Appendix: Guide Questions for the Group Discussion During the Journal Club

Guide questions for group discussion:

1. What are the key terms in this article? How did the authors define them?
2. What is the purported 'gap' in the literature that this paper tried to address?
3. What theories have been considered? What is the underpinning theory used in this article?
4. Identify one or two arguments from the authors.
5. How did they collect research data? Give an example how the authors justify a methods-related decision.
6. What did they do to convince the readers of the validity/trustworthiness/credibility of their findings?
7. How do their findings offer new insight?
8. Identify a phrase that is worth citing or quoting as it may have some relevance to your research.
9. What is the contribution of this study?
10. Identify examples of study limitations.

References

Alguire, P. C. (1998). A review of journal clubs in postgraduate medical education. *Journal of General Internal Medicine, 13*, 347–353.

Bandura, A. (1978). The self system in reciprocal determinism. *American Psychologist, 33*(4), 344–358. https://doi.org/10.1037/0003-066X.33.4.344

Bandura, A. (2001). Social cognitive theory: An agentic perspective. *Annual Review of Psychology, 52*(1), 1–26. https://doi.org/10.1146/annurev.psych.52.1.1

Bochner, A. P., & Ellis, C. (2022). Why autoethnography? *Social Work & Social Sciences Review, 23*(2), 8–18. https://doi.org/10.1921/swssr.v23i2.2027

Braun, V., & Clarke, V. (2021a). One size fits all? What counts as quality practice in (reflexive) thematic analysis? *Qualitative Research in Psychology, 18*(3), 328–352. https://doi.org/10.1080/14780887.2020.1769238

Braun, V., & Clarke, V. (2021b). Can I use TA? Should I use TA? Should I not use TA? Comparing reflexive thematic analysis and other pattern-based qualitative analytic approaches. *Counselling and Psychotherapy Research, 21*(1), 37–47. https://doi.org/10.1002/capr.12360

Elliot, D. L. (2023). *Navigating your international doctoral experience (and beyond)*. Routledge.

Elliot, D. L., Bengtsen, S. S. E., Guccione, K., & Kobayashi, S. (2020). *The Hidden Curriculum in Doctoral Education*. Palgrave Pivot. https://doi.org/10.1007/978-3-030-41497-9

Elliot, D. L., & Makara, K. A. (2021). An online community of international scholars: Enabling spaces for reciprocal academic and psychological support. *Innovations in Education and Teaching International, 58*(6), 693–703. https://doi.org/10.1080/14703297.2021.1991424

Ellis, C., & Bochner, A. (2000). Autoethnography, personal narrative, reflexivity: Researcher as subject. In N. K. Denzin & Y. S. Lincoln (Eds.), *Handbook of Qualitative Research* (2nd ed., pp. 733–768). Sage Publications.

Good, D. J., & McIntyre, C. M. (2015). Use of journal clubs within senior Capstone courses: Analysis of perceived gains in reviewing scientific literature. *Journal of Nutrition Education and Behavior, 47*(5), 477–479. https://doi.org/10.1016/j.jneb.2015.05.003

Harding, J. (2019). *Qualitative data analysis: From start to finish* (2nd ed.).

Harris, J., Kearley, K., Heneghan, C., Meats, E., Roberts, N., Perera, R., & Kearley-Shiers, K. (2011). Are journal clubs effective in supporting evidence-based decision making? A systematic review. BEME Guide No. 16. *Medical Teacher, 33*(1), 9–23. https://doi.org/10.3109/0142159X.2011.530321

Honey, C. P., & Baker, J. A. (2011). Exploring the impact of journal clubs: A systematic review. *Nurse Education Today, 31*(8), 825–831. https://doi.org/10.1016/j.nedt.2010.12.020

Li, S., Malin, J. R., & Hackman, D. G. (2018). Mentoring supports and mentoring across difference: Insights from mentees. *Mentoring & Tutoring: Partnership in Learning, 26*(5), 563–584. https://doi.org/10.1080/13611267.2018.1561020

Newswander, L. K., & Borrego, M. (2009). Using journal clubs to cultivate a community of practice at the graduate level. *European Journal of Engineering Education, 34*(6), 561–571. https://doi.org/10.1080/03043790903202959

Sidorov, J. (1995). How are internal medicine residency journal clubs organized, and what makes them successful? *Archives of Internal Medicine, 155*(11), 1193–1197.

Swift, G. (2004). How to make journal clubs interesting. *Advances in Psychiatric Treatment, 10*(1), 67–72. https://doi.org/10.1192/apt.10.1.67

Wall, S. (2006). An autoethnography on learning about autoethnography. *International Journal of Qualitative Methods, 5*(2), 146–160. https://doi.org/10.1177/160940690600500205

PART II

Insights from Doctoral Supervisors

Facilitating Researcher Independence Through Supervision as Dialogue

Gina Wisker

A doctoral researcher's successful completion of a research project and production of a thesis, monograph or exegesis and publications is usually considered the pinnacle of success for both graduate and supervisor(s). In the most successful relationships, a main aim is fostering both interdependence and the doctoral researcher's confident, enacted, continued independence. For the developing doctoral researcher, successful supervision relationships and processes help establish firm foundations for future research writing and intellectual engagement. Here we consider supervision interactions, including feedback as a developmental dialogue. We share the hidden curriculum practice of developing ideas, arguments and writing together and with others during and beyond the supervision relationship, termed 'Fridaying', (a term historically produced by a participant in one of my

G. Wisker (✉)
University of Bath, Bath, UK

SAARCHI chair, Education, affiliate, University of Johannesburg, Johannesburg, South Africa
e-mail: gw647@bath.ac.uk

D. L. Elliot et al. (eds.), *Developing Researcher Independence Through the Hidden Curriculum*,
https://doi.org/10.1007/978-3-031-42875-3_7

supervisor development workshops) indicating its creativity and equality in a liminal space. (See also Peseta et al.'s chapter on #thesisthinkers.)

In exploring supervisor and doctoral researcher interactions as forms of dynamic developmental dialogues, we can demystify processes and dispel some of the 'fuzziness' surrounding the conceptualisation, development and practices of supervisor/researcher interdependence which respect difference and nurture confidence, autonomy and independence.

Enabling, Demystifying, Nurturing Difference: Supervisory Interactions as Successful Dialogues

Not all doctoral researcher and supervisor relationships are effortlessly enabling with clear communication and increasing successful researcher independence. There can be blockages to all forms of communication based on misunderstandings about responsibilities, to expectations of replication of previous (positive or negative) experiences of interactions, played out in this new supervisory interaction. Many issues around the relationship, the doctoral learning journey, research and the thesis building come from lack of clarity about the goals and shape of the developing research and thesis (or exegesis and articles). One main supervisor role is working with the doctoral researcher to demystify both the process and the product so they can develop independence and ownership of the research journey and writing, fostering 'competent autonomy' rather than 'benign neglect' (Gurr, 2001, p. 85) where (reading the latter generously) ostensible investment in researcher independence leads to a totally hands-off relationship with long communication gaps, when different interactions might actually support progress.

When the supervisory relationship works well it resembles a dance in which each matches responses to the other's aims, needs and steps in the doctoral learning journey (Wisker, 1999). It is also a dialogue in which power and knowledge shift over time to equality. This dialogue is the main developmental interaction on the intellectual and personal learning journey of a doctorate and one in which each party learns, since each researcher, project and supervisor is different, and the challenges and delight of this valuable newness ultimately leads to contributions to knowledge recognised by the achievement of a doctorate. This interdependent, ongoing, positive two-way intellectual dialogue is enabled through knowledge sharing and building and through mutual respect for different needs, understandings and nuances of expression leading to agreement and action. There are structures to work within and towards, and in supervisory interactions, these include learning to work together with appropriate boundaries, duty of care and towards achieving the research learning journey milestones: proposals, ethical clearance, supervisory team meetings, yearly audits, confirmation and transfer, submission, mock viva, examination including viva, and finalising post-examination corrections. Together, you co-develop the research processes for constructing the research design, producing the data, finalising the written work and possibly some co-publishing.

When considering the doctoral journey, there are differences to take into account so supervisors need to adjust interactions in relation to discipline, cultural inflections, modes of study, i.e. part-time, full-time, cohort and individual, part taught course part thesis professional doctorates (DBAs, Prof Docs, EdDs etc.) each of which nuance supervisor/doctoral researcher dynamics, interactions, the pace and kind of work. What each relationship version can encourage and enable is supervisor/researcher interdependence as developing equals, and a confident, competent, flexible, independent researcher who can work beyond the doctorate on a variety of projects. Within the mutually agreed and understood structures of supervisory interactions, institutional and doctoral expectations, there is room and encouragement for individuality, creativity and the extra magic which comes with mutual understanding and respect, building positively upon such differences. In doctoral research learning, knowledge creation and supervisory interactions, these differences of ontology, epistemology, modes of understanding, knowledge construction and expression offer rich, fluid, transformational mutual learning opportunities first realised through dynamic, developmental dialogues founded on the respect between intellectual equals: supervisors and doctoral researchers.

Ecology of Interactions: Supervisor and Doctoral Researcher Dialogues

This piece takes an ecological approach in supporting researcher independence in doctoral education (Elliot et al., 2020) through three forms of dialogue interaction nudged on by supervisors. Each encourages and enables a thriving interdependence than independence within and outside the university to benefit doctoral researchers, supervisors and communities. The aim is demystifying intentions and practices of these interactive processes—where doctoral researchers, supervisors and both formal and hidden curriculum elements interact.

Doctoral researchers engage in many ways in interactive research-oriented dialogues with supervisors, their community of peers, with colleagues in cohorts, at conferences, through emails and through supporting others' writing, or co-writing. Interactive dialoguing with supervisors can enable student autonomy and independence:

1. Structured supervisor/doctoral researcher dialogues held together by a supervisory rhythm from a social start and close, with complex, cognitive, clarifying, contestatory debate, building doctoral-level academic engagement and articulation skills.
2. Structured versions of feedback dialogue.
3. Creative, intellectual, free-flowing 'Fridaying' where intellectual equals co-create in a liminal free space in planned or serendipitous dialogue between doctoral researchers, supervisors, supervisor contacts, colleagues and peers.

Structured Supervisor and Doctoral Researcher Dialogues in Supervision

Some supervisory dialogues are face-to-face while (during Covid and beyond) some are entirely online, synchronous or asynchronous (the latter through email discussion). It could be more difficult to judge responses in online interaction, especially if there is no history of face-to-face interactions (Kumar & Wisker eds., 2022; Wisker et al., 2021). It is, however, always important to think carefully about issues affecting interactions and dialogues, including the culturally inflected learning backgrounds of supervisor and doctoral researcher; developmental dialogue norms such as provocations to query, contest, consider and discuss; politeness and courteous disinclinations to be critical. These sensitivities might be more scripted and difficult online (or less so).

Historically, we (Wisker et al., 2003) conducted research involving supervisor and doctoral researcher interactions based on a counselling model, John Heron's six category intervention analysis (Heron, 1975), as we believed supervision, intended to be enabling, resembling counselling. No team supervisions were included (rare then though now the norm). Their dynamic could produce quite different interactions and research would be interesting.

Following ethical approval (Anglia Ruskin University, 2001), we asked supervisors and students to consider taping a supervision interaction which we analysed and shared with participants before publication. No one redacted their transcripts.

Working with the Dialogues

The range of supervisory dialogue types developed during our research prompt questions about kinds of interactions, why, to what effects and how could they help move doctoral researchers on and empower them.

In supervisor workshops I ask: Could you use these kinds of interactions? Why? When? Why not? Are there any others you might prefer to use? The intention is to consciously surface how we engage and why, to what outcomes and to prompt development of an effective repertoire.

Textbox 1 Supervisory dialogues

1. didactic
2. prescriptive
3. informative
4. confronting (or challenging)
5. tension relieving/social
6. encouraging critical, problematising and problem-solving attitude
7. eliciting
8. supporting
9. summarising
10. clarifying
11. collegial exchange (*Wisker et al., 2003, after* Heron, 1975)

Through thematic analysis of supervisory dialogues, we found ten questioning themes/interactions or 'intervention' categories, developing from Heron's 'six category intervention analysis' (1975). Supervisors might have preferences and also need to determine the language appropriate for each category of interaction, where and when to use them to enable the doctoral researcher with whom they work.

More collegial and social interactions start and end supervisions or release tension in fraught or focused discussions. Some dialogues are informational or instructional. Others engage contrasting and critical thinking, ways of working with conflicting information, nudging more complex theorising, relating theory to emerging data.

In other interactions, supervisors 'nudge' doctoral researchers to move further, more deeply into their work. There are also student response themes which emerged (2003), less refined as yet. It is useful to consider these regarding what might shut doctoral researchers down, and what might enable and empower their interactions, confidence, articulation, critical and other thinking.

Textbox 2 Student (doctoral researcher) response themes (still under development)

1. seeking direction and information;
2. seeking feedback;
3. information giving;
4. information seeking;
5. working out through talk/developing ideas and plans through dialogue;
6. student defining ideas;
7. student developing ideas;
8. student judgement re: needs;
9. student pleasing supervisor;
10. student relating previous work to own work, theory to practice, experience to research culture;
11. student taking control;
12. tentative-provisional thinking;
13. uncertainty (of reaching PhD) unclear end result;
14. clear idea of the project as PhD. (Wisker et al., 2003)

Interestingly, responses evidence development of both independence and confidence shown in examples of moving from more tentative questioning, forming of ideas to clarity about the PhD project.

In supervisory development workshops, these dialogues help us consider how different doctoral researcher responses might reflect different needs and development along the doctoral learning journey; how to work with difference, encourage and enable more empowered, independent, appropriately confident responses which develop over time, but for some never develop and for others are evident early on—perhaps with professionals successful in another context. The dialogues can help us consider and work with different researcher needs, at different stages from proposals through research and writing to submission and beyond. They suggest the shape and flow of supervisions for all involved. Seeing these two sets of responses we are better placed to plan and choose forms of interaction and wording, encouraging enabling response and work with facilitative dialogue.

The most useful dialogues take place where both participants match cognitive processes and move forwards, so doctoral researchers increasingly take the lead.

Some early dialogues focus on functional and conceptual work, setting up ways of working together, developing research proposals and ethical approvals (if appropriate) and creating a conceptual framework informing research. While functional interactions can be rather one way, i.e. instructive and informational, they can also be conducted through discussion, questioning and prompting, leading doctoral scholars to understand early seemingly only bureaucratic processes as times for thinking through theory, big ideas and ways to undertake their research. Working relationships established encourage doctoral researchers to explore their own ideas in their own terms, refine and shape these into doable, complex enough projects with structure and scaffolding. Some see the doctorate as a lifelong project (Mullins & Kiley, 2002) but in the event it must be manageable, finished within a reasonable timeframe. Boundaries, limitations and pragmatic choices are important.

Pragmatic functional dialogues help doctoral researchers trim, hone and make realisable their projects and the writing about them, while more exploratory, conceptual, questioning, problematising dialogues encourage theorising and free-flowing creative thinking. (*This is linked to Frick's chapter discussing how supervisors can unlock and stimulate doctoral scholars' independence and creativity by employing pedagogical strategies.) Following doctoral scholars, I and colleagues explored dialogues towards completion, noting 'Dialogues aim to encourage developing the thesis and preparing for the viva. Students are asked to indicate crucial change moments of their research ... Facing up to and identifying the effects of critical incidents moves learners on in their ownership of this learning' (Wisker et al., 2003).

Over time, we engage supervision modes to guide, prescribe, inform, confront, elicit, clarify, support, summarise and move the work on. There will be moments of 'learning leaps', blockages, disagreements and clarification where doctoral scholars recognise they must maintain momentum, fulfil requirements, that decisions, and progress are dependent upon their ability to problem-solve, make choices, take risks, be original. Through learning conversations/supervisory dialogues, doctoral scholars recognise for themselves where to pull ideas and information together into a synthesis, engaging theoretical perspectives and critical reading in a dynamic dialogue with their own work.

Feedback as Developmental Dialogue (Wisker, Online)

Much feedback focuses on correcting errors, offering information, while feedback or feedforward (Race, online) aimed at encouraging further learning also supports further, new learning, empowering doctoral researchers to identify, manage their own issues and make their own decisions. Rowena Murray's feedback characteristics helpfully identify the feedback range offered in doctoral level writing. Concerned about explicit guidelines in supervisor feedback from the student's point of view, Murray (2002) asks:

> Are the comments global or detailed or both? For supervisors, there is a decision to make about what type of feedback to give. Do they want to make you focus on the 'big picture' of your whole argument, or a section of it? Or do they want you to tidy up the style? Is clarification of terms paramount? … supervisors may recognize that one is more important, at this stage, than the others. (Murray, 2002, p. 78)

Supervisors comment on elements of achievement or need in the doctoral scholars' work, but it would overload if we did that every time. Developing their own engagement with feedback, processing, owning and acting on it helps establish independence through internal interaction and enables them to internalise corrective and developmental learning which much feedback intends, fully owning their own decisions.

Kumar and Stracke developed a framework of useful feedback categories (2007, p. 465) based on speech functions helping supervisors define what to say, where and when, for specific outcomes.

They consider feedback responses and their intent as: Directive, Expressive and Referential.

Directive—corrects errors and informs, '48% not 54%'; 'Give full reference for Vygotsky'.

Expressive—comments on the quality of the work, offering praise or criticism, directly or sometimes indirectly: 'this is developing well', 'clearly expressed' or 'needs further clarity—what do you mean by xxx???'

Referential provides information: 'you need a fuller reference here' also offering explanations, models and fuller guidance, e.g. 'Look at this example of using XX's theory on the relationship between place and identity. ... How might you use xxx's theory to engage with your ideas and arguments about x??'. The latter example involves doctoral researchers thinking through a model, not just copying it, working out how it might inform their own work, active in responding to learning suggestions. If we want doctoral researchers to take ownership and become independent, they should correct errors but also be nudged into thinking, learning from examples and models, appreciating and deciding on their arguments in a contested situation. This makes the interaction a dialogue encouraging independent thought, rather than an instruction shutting it down.

As supervisors, we consider how and when to use categories of feedback in our own work, constructing balanced feedback appealing to a variety of doctoral scholars. It is not a dialogue, of course, unless the doctoral scholar takes the suggestions and prompts into their own thinking and practice.

Ashtarian and Weisi (2016) note affective language, e.g. 'please' and 'could you', politely preventing the reading of feedback as critical or destructive, while Parr and Timperley (2010) advise that suggesting improvements to approach a desired response needs careful management as it could look like pointing out lack, shutting the researcher down. This is a complex linguistic maze to negotiate, particularly if some of the expressions, the politenesses and the shortcuts of '???', 'say more', 'clarify' do not immediately speak to any previous comments received. For many doctoral researchers, the language of the research and thesis might be their second, third, fourth, etc., and both sensitivity and accuracy are needed in translation of tone, content, advice to empower and enable them to interpret, own and act on feedback. Building on limited work on supervision feedback on doctoral work with second-language speakers (L2), in her own work on supervisor feedback with Chinese doctoral researchers in New Zealand, LinLin Xu (Xu & Grant, 2017; Yu & Lee, L. (2016) used Bakhtin's dialogic theories to explore the dialogue, present, or not, in such feedback interactions, noting that supervisor comments ranged from informative and corrective to engaging in a more empowering discussion. Some (Yu & Lee, 2016) suggest doctoral scholars might benefit from scaffolding, 'progressive development' in feedback comments, suggesting that students appreciate scaffolding and comments, offering ways they could improve 'rather than a simple judgement without explanation' (Xu & Grant, 2017, p. 21).

Bearing language issues and potential misinterpretations in mind, I argue that clear comments and a dialogue which helps structure future work, as well as encouraging critical thinking, ownership of the work and independence, are useful for all doctoral researchers.

Fridaying Dialogues

Not all supervision and research-related dialogues are formal whether in supervisions or written feedback. Much development of thought, understanding and creative thinking emerges as unscripted through dialogues which are part of the hidden curriculum. 'Fridaying' (a term established by a South African colleague in one of my suprvisor development workshops, and his writing collaborator, on which I build here) uses planned or unplanned gaps and relaxing moments to engage in an intellectual creative space and co-build complex, shared thinking and dialogue, leading to mutual understanding of research decisions and achievements and often to co-writing. It takes place freely in the interstices of more structured interactions, such as conferences, either planned or serendipitously, and makes the most of often hard-won free space and time. Anyone interested can be involved usually including colleagues, co-researchers (or future co-researchers/co-authors) and supervisors with doctoral researchers (not always their 'own' students). My colleague Gillian Robinson and I recognised the term as defining our sudden clever thoughts on the outskirts of planned conference moments (which had to be talked through, taped, worked with), as out-takes when running supervision workshops abroad on a large cohort-based doctoral programme, and as discussions with others in quiet reflective moments. Fridaying, as I interpret it, is a creative dialogue suddenly ignited by a shared sparky thought or an ongoing set of actions and reflections between equals. It involves one complex thought springing from the other, recorded, with permission, and then used in research. We spot its beginnings as we talk with each other and with doctoral scholars around residential courses, on walks, over coffee or supper.

The 'Friday' element is the liminal free space in which igniting of ideas and developing discussion begin and then are 'nudged' to open up thinking. It can lead to learning leaps, breakthroughs in understanding or 'conceptual threshold crossings' (Kiley & Wisker, 2009). Work on dark and light sides of supervision (Wisker et al., 2017) came from this process. Setting up moments when it is likely to start (co-attending conferences, having a coffee and chatting) and fostering the energies with doctoral researchers is something supervisors can do to encourage independence. (See also Makara et al.'s chapter discussing participation in a Journal Club to elucidate the Fridaying concept.)

Conclusion: What Do Supervisors Do to Support Independence?

Supervisors work with very different research learners, adjusting support to different needs, not substituting the supervisor's practices for those the researcher should develop themselves, rather offering examples, nudging opportunities for new research practices, different forms of expression. Supervisors demystify and enable further entrance into the worlds of research and publication, which are otherwise likely to remain esoteric and obscure for researchers from all contexts whether related to discipline, mode of study, gender, international, culture, class or learning behaviour. Supervisors open doors to and illuminate elements of the hidden curriculum of research-related behaviours and participants, enabling researchers to find their own guided way through the complexities of the research planning, decisions, activities, overcoming hurdles in writing, examination submission and publication. They introduce researchers to communities of other researchers working internationally, to other projects and to publishing politics and practices. I here emphasise the opening of doors, demystification, modelling, sharing and networking enabled by interactive dialogues, dynamic interactions and practical actions. Crucially, interdependence and independence-oriented supervisor guidance helps researchers develop the independence and insights to continue to work through and beyond achieving the doctorate.

Using research and experience on and with supervisors and doctoral researchers, this chapter explores ways in which supervisory dialogues, constructive feedback/feedforward, modelling, networking, community-building and sharing of good practice induct researchers into self-aware, reflective, successful independence during the doctoral research process, enabling continued interdependence and independence, leading to future community building, researcher development, leadership and, probably, if appropriate, effective dialogue-based supervisory practices.

References

Ashtarian, S., & Weisi, H. (2016). A comparative study of the use of affective markers in supervision comments: Case of native versus non-native supervision comments. *RELC Journal, 47*(2). https://doi.org/10.1177/0033688216631172

Elliot, D. L., Bengtsen, S. S. E., Guccione, K., & Kobayashi, S. (2020). *The hidden curriculum in doctoral education.* Palgrave Macmillan.

Gurr, G. (2001). Negotiating the "Rackety Bridge" – A dynamic model for aligning supervisory style with research student development. *Higher Education Research and Development, 20*(1), 81–2 (85). https://doi.org/10.1080/07924360120043882

Heron, J. (1975). *Six category intervention analysis.* Tavistock Institute.

Kiley, M., & Wisker, G. (2009). Threshold concepts in research education and evidence of threshold crossing. *Higher Education Research & Development, 28*, 431–434.

Kumar, S., & Wisker, G. (eds.). (2022). *Innovations in education and teaching international, special issue on supervision online.*

Kumar, V., & Stracke, E. (2007). An analysis of written feedback on a PhD thesis. *Teaching in Higher Education, 12*(4), 461–470.

Mullins, G., & Kiley, M. (2002). It's a Ph.D., not a Nobel Prize How experienced examiners assess research theses. *Studies in Higher Education, 27*(4), 369–386. https://doi.org/10.1080/0307507022000011507

Murray, R. (2002). *How to write a thesis.* Open University Press.

Parr, J., & Timperley, H. (2010). Feedback to writing, assessment for teaching and learning and student progress. *Assessing Writing, 15*(2), 68–68.

Wisker, G. (1999). Learning conceptions and strategies of postgraduate students (Israeli PhD students) and some steps towards encouraging and enabling their learning: 2nd Postgraduate Experience Conference: Developing Research Capacity in the New South Africa, Cape Town, South Africa.

Wisker, G., McGinn, M. K., Bengtsen, S. S. E., Lokhtina, I., He, F., Cornér, S., Leshem, S., Inouye, K., & Löfström, E. (2021). Remote doctoral supervision experiences: Challenges and affordances. *Innovations in Education and Teaching International, 58*(6), 612–623. https://doi.org/10.1080/14703297.2021.1991427

Wisker, G., Robinson, G., & Bengtsen, S. (2017). Penumbra: Doctoral support as drama: From the 'lightside' to the 'darkside.' From front of house to trapdoors and recesses. *Innovations in Education and Teaching International, 54*(6), 527–538.

Wisker, G., Robinson, G., Trafford, V., Warnes, M., & Creighton, E. (2003). From supervisory dialogues to successful PhDs: Strategies supporting and enabling the learning conversations of staff and students at postgraduate level. *Teaching in Higher Education, 8*, 383–397. https://doi.org/10.1080/13562510309400

Xu, L., & Grant, B. (2017). International doctoral students' becoming: A dialogic perspective. *Innovations in Education and Teaching International, 54*, 1–10. https://doi.org/10.1080/14703297.2017.1318711

Yu, S., & Lee, I. (2016). Peer feedback in second language writing (2005–2014). *Language Teaching, 49*(4), 461–493. https://doi.org/10.1017/S0261444816000161

Websites

Race, P. https://phil-race.co.uk/2018/02/feedback-feedforward-just-tips-including-students/.

Wisker, G. The good supervisor. https://goodsupervisor.co.uk/ (login: brighton).

Creative Supervising/Supervising for Creativity: Exploring the Hidden Dimensions of Creativity in Doctoral Supervision

Liezel Frick

Even though creativity is implicitly expected at the doctoral level, it often remains a hidden aspect of doctoral curricula across disciplines and doctoral systems. As such, it often forms part of what Elliot et al. (2020, p. 3) refer to as "the unofficial (and informal) channels of genuine and useful learning", the so-called ***hidden curriculum*** in doctoral education. Yet, doctoral creativity underlies the generic expected outcomes of any doctorate—independent and original knowledge creation.

L. Frick (✉)
Centre for Higher and Adult Education, Stellenbosch University, Stellenbosch, South Africa
e-mail: blf@sun.ac.za

D. L. Elliot et al. (eds.), *Developing Researcher Independence Through the Hidden Curriculum*,
https://doi.org/10.1007/978-3-031-42875-3_8

Supervisors play a key role in unlocking and stimulating doctoral scholars' in-/inter-dependence[1] and creativity and employ pedagogical strategies in order to do so. This mini-chapter sets out to conceptualise what creativity means in the context of doctoral supervision and highlight the difference between creative supervising and supervising for creativity as different pedagogical strategies towards developing scholarly in-/inter-dependence.

Demystifying Doctoral Creativity

Doctoral creativity can be defined as a *process* focused on (research) work that is both novel and relevant to a particular socio-cultural-historical context—such as an academic discipline—that realises and expresses the true potential of both the person and the product/outcome (Baptista et al., 2015; Frick, 2012).

Wallas's (1926) cyclical model of the creative process encompassing five phases (including preparation, problem-solving, incubation, moment of insight, and verification and application) highlights that creativity does not emerge suddenly but needs to develop and be fostered over time in an atmosphere that allows exploration and expression, regardless of the discipline or programme format (Jones, 1972). Incubation, in particular, is probably the most hidden part of the creative doctoral process, often underrated, and therefore left unaccounted for in the doctoral curriculum. Students may furthermore erroneously equate incubation and procrastination. Their perplexity over the difference between these states of engagement may be exacerbated by institutional demands for timely completion and evidence of originality. Despite it being difficult to determine the role supervisors play in the incubation phase towards stimulating creativity, Eva Brodin and I (2011) postulated that supervision (and, therefore, doctoral pedagogy) plays a crucial role in developing creative capabilities of doctoral scholars.

A pedagogical understanding of creativity at the doctoral level demands a nuanced appreciation of the interplay between the inherent qualities of

[1] Even though independence is an envisioned generic doctoral outcome, an argument could be made that knowledge creation is never truly independent, but rather inter-dependent. All researchers engage with foundations of existing knowledge and are thus inter-dependent (rather than independent) to their scholarly communities. Doctoral supervisors stand at the interface of this inter-dependence, acting as intermediaries between the scholarly community and the doctoral scholar.

both doctoral scholars and their supervisors, supervisory practices, and environmental factors that interact in the process of doctoral becoming and independence. If the pedagogical approach does not explicitly facilitate and value creativity, it is unlikely that the expected doctoral in-/inter-dependence will manifest. There is, therefore, value in demystifying doctoral creativity and considering the role of supervisors in enabling such creativity (Brodin & Frick, 2011).

The Creative Pedagogue: Creative Supervising

Much of the current research around doctoral creativity has focused on developing doctoral scholars' creativity (see, e.g., Bengtsen, 2017; Brodin, 2016, 2018; Frick & Brodin, 2020; Ulibarri et al., 2014). While intrinsic, task-focused motivation seems an important prerequisite for creativity (Dewett et al., 2005), external motivators—including supervisors—may also play a role in enabling creativity (Gardner, 1988). The role of external motivation leads us to consider a supportive and rewarding environment (and integral to this environment, the role of the supervisor) as a necessary stimulus for creativity. Although we know that supervisors play a determining role in doctoral progress and ultimate success (Sverdlik et al., 2018; Young et al., 2019), and their roles and responsibilities have been the focus of multiple studies (see, e.g., Gatfield, 2005; Halse & Malfroy, 2010; Lee, 2008), developing their own creativity in stimulating doctoral scholars' creativity has remained largely hidden from scholarly view.

Pedagogy is a formative element in doctoral becoming and may be a catalyst or inhibitor for creativity. Through pedagogy students become socialised into the academic community (McWilliam et al., 2008), which provides a sense of collective direction (McWilliam & Dawson, 2008). Conceptualising doctoral pedagogies along these lines provides a useful point of departure to consider doctoral creativity not only as something that needs to be enacted by doctoral scholars alone but rather as an intricate inter-dependent interplay of co-creation between doctoral scholars *and* their supervisors (even though the creative role of the supervisor often remains hidden)—what Elliot et al. (2020) refer to as an expanded pedagogy.

Lee and Danby (2012) argue that pedagogical practices need to be made explicit and visible if they are to be of any use to others. Pedagogy lies at heart of doctoral education if we use Lusted's (1986, pp. 2–3) conceptualisation wherein pedagogy refers to the relationship between the learner (the doctoral scholar), the teacher (in this case the supervisor(s)), and the knowledge *created* within this relationship (the so-called original contribution). According to Howard and Turner-Nash (2011, p. 23), pedagogy is a process where meaning is constantly (re-)created and where the identities of those involved develop through discursive practices and power/knowledge relations. To this end, critical pedagogy serves the purpose of doctoral education in that it offers all the role players agentic entry into the *shared pursuit of knowledge*—highlighting the inter-dependent nature of doctoral education. Doctoral pedagogy, thus, involves the knowers (doctoral scholars and supervisors alike), the known and the unknown, and what the rules of engagement are under which these elements combine to eventually create knowledge—the ultimate goal of a doctorate. (See also Wisker's chapter on supervisors employing pedagogical strategies to stimulate doctoral scholars' creativity.)

What does it then mean to be a creative pedagogue? Even though this notion remains largely hidden in literature on "good" supervision, there are some hints we can take forward towards enabling supervisors to become co-creators of knowledge more explicitly.

Creative Pedagogues Are Experts at Play

Creativity is not possible without a thorough understanding of the basic principles of and knowledge within a field of study. The importance of knowledge and immersion in the field of study in identifying problems and gaps in order to move beyond the existing perspectives and to create something new is well recognised (Dewett et al., 2005). Creativity results from purposeful behaviour, and often lengthy and arduous processes (Sternberg & Lubart, 1999), but which Pope (2005, p. 31) still describes as "work at play". Being able to use foundational knowledge and expertise in developing a research idea, challenging existing understandings, and eventually extending a field of research demand certain level of creative playfulness. Creative pedagogues in the context of doctoral education are, therefore, active researchers themselves, who work at the frontiers of their fields of study and are able to draw on their expertise to challenge the boundaries of these fields.

Creative Pedagogues Have a Tolerance for Ambiguity and Allow for Constructive Friction

Developing your own creativity is difficult. Facilitating this development in others is even more so. Creative scholars are often non-conformists, which may result in tension and adjustment problems (Jones, 1972). Such scholars often strive for independence, are curious and perceptive, search widely for related information, act intuitively, do not like being confined to pre-determined courses, and need to explore options—even though some options may lead to failure. In addition, not all such scholars will develop in similar ways, or in a linear fashion, or to the same level of manifestation equally in all the research phases which requires a tolerance of ambiguity as there is no set course or direction that would enable such creative endeavours. The ideal learning environment for these scholars would permit what Vermunt and Verloop (1999) call constructive friction, thus understanding that different viewpoints could lead to creative advances if the learning environment creates safe spaces for exploration and experimentation. Such creative pedagogues have (and instil) creative confidence. Creative confidence entails having the space and willingness to take creative risks and fail but knowing that each idea created has value. Creative pedagogues who can instil creative confidence in doctoral scholars enable them to become creative researchers themselves throughout their research careers (Ulibarri et al., 2014).

Creative Pedagogues Have (Research) Integrity

Research is risky (Frick et al., 2014), but such risk is necessary for enabling creativity. While acknowledging the inherent risks of research may help to avoid hazardous risk, it is sometimes difficult for supervisors to manage cutting-edge creative research adaptively and responsibly, especially amid institutional systems that have become highly bureaucratised and risk averse. However, creativity demands a certain tolerance of risk while knowing that we also need to question and problematize creativity as inherently "good". There is a tension in achieving this kind of balance which is often hidden in doctoral curricula. Creative pedagogues understand that creativity has consequences. Being creative raises serious ethical issues, including possibly breaking rules and standard operating procedures, challenging authority and avoiding tradition, creating conflict, competition and stress, and taking risks (Baucus et al., 2008). But creative pedagogues maintaining research integrity throughout navigating and balancing research risk and creativity (Frick, 2021).

Creative Pedagogues Are Radical Collaborators

Creativity is often not the result of individual endeavour alone (Csikszentmihalyi, 1999). The greater problems facing science and society demand collaboration across disciplines (Manathunga et al., 2006), even though impermeable discipline-based boundaries between disciplines are still evident in much of the doctoral research produced worldwide. Ulibarri et al. (2014) refer to the intentional creation of diverse teams to solve problems as radical collaboration. Disciplinary boundary crossing promotes higher-order thinking, an understanding of divergent knowledge systems, and creative problem-solving behaviour. In addition, much of this kind of work remains hidden, un-recognised, and unrewarded. In addition, these notions often run counter to the apprenticeship model of supervision that is still evident in some disciplines and settings (Mkhize, 2022; Lee & Bongaardt, 2021). Creative pedagogues are those who find ways of transcending such boundaries and create spaces for doctoral scholars to do the same.

Unlocking What Is Hidden: Supervising for Creativity

While supervisors themselves may be creative and infuse that creativity into their work with doctoral scholars, *how* they unlock doctoral scholars' creative potential often remains a hidden part of the doctoral curriculum. MacKinnon (1970) warns that creativity should not be seen as something to be taught, but rather as developed by leading through example. Since creativity is both an innate and a learned quality, it can be developed, even though an innate talent for a specific field is, of course, helpful (McWilliam & Dawson, 2008). Supervision is key to unlocking this creative potential in doctoral scholars—thus supervising *for* creativity.

Supervising for creativity differs from being a creative pedagogue in that the latter speaks to the identity of the supervisor (thus *being* creative), while the earlier relates to pedagogical practices that would enable creativity in others (thus *doing* supervision in a way that enables creativity). From a pedagogical point of view, how do we enable both the individual and the individual as part of a group to become creative? I would argue that in this case, being could enable doing—being a creative pedagogue could enable supervising for creativity. I, therefore, build on the earlier notion of being a creative pedagogue in considering what supervising for creativity might mean.

Supervision for Enabling Scientific Play

Scientific playfulness may not come naturally to doctoral scholars. By involving doctoral scholars in all the phases of their own research—from conceptualisation and planning to eventual reporting—supervisors can enhance doctoral scholars' creativity interdependently. If doctoral scholars are able to draw from the examples of creativity set by their supervisors, their own creative development becomes more likely. Some examples of pedagogic practices that could enable scientific play include providing early and frequent feedback to doctoral scholars' research ideas, collaboratively generating multiple research questions and/or hypotheses, and providing input on which questions are worth pursuing, requiring doctoral scholars to transfer knowledge from one area to another, search for common principles where facts from different areas of knowledge can be related, and engage in imaginative experimentation. In this way, supervisors can help these scholars to step back from facts to gain a greater perspective.

Supervision Amidst Ambiguity that Allows for Constructive Friction

A basic scientific premise is *doubt*. Yet, in order to be constructively doubtful, in, for example, coming up with a hypothesis, a scholar needs to build such a hypothesis on a set of assumptions that need to provide some certainty. Within this interplay between certainty and doubt lies a supervisory paradox. Supervisors who are able to create a space for debate through problematising and deconstructing knowledge, and who promote a respectful yet challenging learning environment, create spaces for creativity to develop. Such spaces are dependent upon nurturing, student-centred learning environments that provide a solid scientific foundation, yet value divergence and diversity. These spaces also make allowances and accept failure as part of the learning process.

Supervision that Promotes Research Integrity

Creative pedagogues understand their responsibility to ensure that doctoral scholars appreciate their moral responsibility to consider the social and ecological consequences of their research, thus encouraging research that explores ways of balancing rather than controlling risk, while encouraging creativity. Envisioning future possible outcomes of research results through scenario planning could help doctoral scholars to imagine the consequences of their research projects and help them tread more carefully in planning and executing research with integrity. Doctoral scholars will often not do this kind of envisioning independently, thus supervisors play an integral role in not only adhering to ethics protocols but instilling a longer-term understanding of responsible scholarship.

Supervision for Radical Collaboration

Transcending disciplinary boundaries may be hard to implement in practice, and creativity may be more difficult when rigid disciplinary boundaries exist, as creativity requires a redefinition of the permissible problems, concepts, and explanations within the discipline and its scientific community. Creating spaces that support exploration across disciplinary boundaries leads to unique challenges for doctoral pedagogy in which creativity needs to be fostered but can be rewarding in facilitating truly creative projects. Doctoral scholars are rarely able to creatively challenge existing ways of thinking and doing in isolation or without some encouragement or examples to follow. Team supervision, co-supervision, and peer learning are some ways in which supervisors may create enabling environments for creativity to develop. Journal clubs, lab meetings, and research group discussions could supplement individual consultations. Those who are more advanced often help those at a less advanced level or those who have started their studies more recently, which lightens the load of the supervisor. Co-publication of research results may also enable such co-becoming. (Some examples of supervising for creativity are presented in chapters by Makara et al., Peseta et al., and Preece.) Another example involves the development of the group dissertation for certain disciplines. These so-called capstone projects not only encourage doctoral scholars to work collaboratively, they often also involve external stakeholders.

Beyond the Hidden Curriculum—Creativity as a Key Supervisory Concern

Even though creativity is deeply rooted within the doctoral curriculum, it lays the foundation for the original contribution all doctoral scholars are expected to eventually make. Yet, such creativity does not develop in isolation or without pedagogical intervention. This chapter explored doctoral creativity from a pedagogical perspective, considering the enablement of creativity as a key supervisory responsibility in terms of both creative supervising and supervising for creativity. As such, it addressed the current gap in literature on doctoral creativity as a supervisory concern.

Creative supervising speaks to what it means to be a supervisor when creativity is foregrounded, while supervising for creativity relates to pedagogical practices that support the development of doctoral creativity. Playfulness, tolerance of ambiguity and constructive friction, research integrity, and radical collaboration were foregrounded as key elements of both being and doing within the context of doctoral supervision.

In-/inter-dependent creativity will not develop within and beyond the doctorate without creative pedagogues who supervise for creativity. This chapter opens a window onto this hidden curricular space.

Declarations of Conflicts of Interest I have no evident conflict of interest to declare

Acknowledgements None

References

Baptista, A., Frick, L., Holley, K., Remmik, M., Tesch, J., & Åkerlind, G. (2015). The doctorate as an original contribution to knowledge: Considering relationships between originality, creativity, and innovation. *Frontline Learning Research, 3*(3), 55–67.

Baucus, M. S., Norton, W. I., Baucus, D. A., & Human, S. E. (2008). Fostering creativity and innovation without encouraging unethical behavior. *Journal of Business Ethics, 81*(1), 97–115.

Bengtsen, S. S. (2017). An exploration of darkness within doctoral education: Creative learning approaches of doctoral students. In *Handbook of research on creative problem-solving skill development in higher education* (pp. 260–282). IGI Global.

Brodin, E. M. (2016). Critical and creative thinking nexus: Learning experiences of doctoral students. *Studies in Higher Education, 41*(6), 971–989.
Brodin, E. M. (2018). The stifling silence around scholarly creativity in doctoral education: Experiences of students and supervisors in four disciplines. *Higher Education, 75*(4), 655–673.
Brodin, E. M., & Frick, B. L. (2011). Conceptualising and encouraging critical creativity in doctoral education. *International Journal for Researcher Development, 2*(2), 133–151.
Csikszentmihalyi, M. (1999). Implications of s systems perspective for the study of creativity. In R. J. Sternberg (Ed.), *Handbook of creativity* (pp. 325–339). Cambridge University Press.
Dewett, T., Shin, S. J., Toh, S. M., & Semadeni, M. (2005). Doctoral student research as a creative endeavour. *College Quarterly, 8*(1), 1–20.
Elliot, D. L., Bengtsen, S. S., Guccione, K., & Kobayashi, S. (2020). *The hidden curriculum in doctoral education.* Springer Nature.
Frick, B. L. (2012). Pedagogies for creativity in science doctorates. In A. Lee & S. Danby (Eds.), *Reshaping doctoral education: Programs, pedagogies, curriculum.* Routledge.
Frick, B. L. (2021). Doctoral creativity as an epistemological force in saving and/or destroying the world. In R. Barnacle & D. Cuthbert (Eds.), *PhD at the end of the world.* Springer.
Frick, B. L., Albertyn, R. M., & Bitzer, E. M. (2014). Conceptualising risk in doctoral education: Navigating boundary tensions. In E. M. Bitzer, R. M. Albertyn, B. L. Frick, B. Grant, & F. Kelly (Eds.), *Candidates, supervisors and institutions: Pushing postgraduate boundaries.* African SunMedia.
Frick, B. L., & Brodin, E. M. (2020). A return to Wonderland: Exploring the links between academic identity development and creativity during doctoral education. *Innovations in Education and Teaching International, 57*(2), 209–219.
Gardner, H. (1988). Creative lives and creative works: A synthetic scientific approach. In R. J. Sternberg (Ed.), *The nature of creativity* (pp. 298–324). Cambridge University Press.
Gatfield, T. (2005). An investigation into PhD supervisory management styles: Development of a dynamic conceptual model and its managerial implications. *Journal of Higher Education Policy and Management, 27*(3), 311–325.
Halse, C., & Malfroy, J. (2010). Retheorizing doctoral supervision as professional work. *Studies in Higher Education, 35*(1), 79–92.
Howard, K. J., & Turner-Nash, K. (2011). Alimentar: Theorizing pedagogy, curriculum, and mentorship for democratic doctoral education. *Equity and Excellence in Education, 44*(1), 22–39.
Jones, T. P. (1972). *Creative learning in perspective.* University of London Press.
Lee, A. (2008). How are doctoral students supervised? Concepts of doctoral research supervision. *Studies in Higher Education, 33*(3), 267–281.

Lee, A., & Bongaardt, R. (2021). Conclusion: The future of doctoral research in the light of experience. In *The Future of Doctoral Research* (pp. 345–353). Routledge.

Lee, A., & Danby, S. (2012). *Reshaping doctoral education: International approaches and pedagogies*. Routledge.

Lusted, D. (1986). Why pedagogy? *Screen, 27*(5), 2–14.

MacKinnon, D. (1970). Creativity: A multi-faceted phenomenon. In J. D. Roslansky (Ed.), *Creativity* (pp. 17–32). North-Holland.

Manathunga, C., Lant, P., & Mellick, G. (2006). Imagining an interdisciplinary doctoral pedagogy. *Teaching in Higher Education, 11*(3), 365–379.

McWilliam, E., & Dawson, S. (2008). Teaching for creativity: Towards sustainable and replicable pedagogical practice. *Higher Education, 56*, 633–643.

McWilliam, E., Poronnik, P., & Taylor, P. G. (2008). Re-designing science pedagogy: Reversing the flight from science. *Journal of Science Education and Technology, 17*, 226–235.

Mkhize, N. (2022). The state of doctoral training in South Africa: Current and emerging practices. In *Doctoral training and higher education in Africa* (pp. 36–60). Routledge.

Pope, R. (2005). *Creativity: Theory, history, practice*. Routledge.

Sternberg, R. J., & Lubart, T. I. (1999). The concept of creativity: Prospects and paradigms. In R. J. Sternberg (Ed.), *Handbook of creativity* (pp. 3–15). Cambridge University Press.

Sverdlik, A., Hall, N. C., McAlpine, L., & Hubbard, K. (2018). The PhD experience: A review of the factors influencing doctoral students' completion, achievement, and well-being. *International Journal of Doctoral Studies, 13*, 361–388.

Ulibarri, N., Cravens, A. E., Cornelius, M., Royalty, A., & Nabergoj, A. S. (2014). Research as design: Developing creative confidence in doctoral students through design thinking. *International Journal of Doctoral Studies, 9*, 249–270.

Vermunt, J. D., & Verloop, N. (1999). Congruence and friction between learning and teaching. *Learning and Instruction, 9*, 257–280.

Wallas, G. (1926). *The art of thought*. Harcourt Brace.

Young, S. N., VanWye, W. R., Schafer, M. A., Robertson, T. A., & Poore, A. V. (2019). Factors affecting PhD student success. *International journal of exercise science, 12*(1), 34.

Developing a Disposition for Harnessing the Hidden Curriculum En Route to Becoming Independent Researchers: The Role of Doctoral Supervisors

Dely Lazarte Elliot

Introduction

Lessons acquired via the hidden curriculum within the context of doctoral education tend to be equally elusive and ubiquitous primarily due to their unintended and unstructured nature. Not only do such lessons take various shapes and forms, but they are also likely to occur in various ecological milieu or nested contexts (Bengtsen & McAlpine, 2022; Elliot et al., 2020; McAlpine & Amundsen, 2018; McAlpine & Norton, 2006). These hidden curricular lessons may arise in doctoral scholars' numerous interactions with fellow doctoral scholars, supervisors, post-doctoral scholars, course, seminar and workshop leaders and participants and many others—both within and outwith the academic context. In turn, along with direct

D. L. Elliot (✉)
University of Glasgow, Glasgow, UK
e-mail: Dely.Elliot@glasgow.ac.uk

D. L. Elliot et al. (eds.), *Developing Researcher Independence Through the Hidden Curriculum*,
https://doi.org/10.1007/978-3-031-42875-3_9

and formal learning (also called 'curriculum proper'), doctoral scholars are able to enrich their experience via unintended complementary learning offered by these non-scholarly activities and through their non-academic counterparts (Martin, 1994).

However, while this type of learning might be in abundance, it does not necessarily guarantee what might be referred to as 'learning consumption' (or utilising learning); instead, genuine learning is often not recognised, let alone realised or used to the scholars' full advantage. Due to its tacit presence, it can at times be tricky to find the hidden curriculum (Elliot, 2023; Elliot et al., 2016). Yet, given that scholars are likely to benefit from the hidden curriculum, this raises the question as to how finding this form of curricular learning can be encouraged. More specifically, how can hidden curriculum learning be harnessed more strategically with a view to supporting all scholars, particularly the doctoral scholar cohort who are en route to becoming independent researchers? These exemplify the questions that prompted my reflection as a supervisor and a staff member who supports doctoral scholars and which I will endeavour to address in this chapter.

Despite its inherent elusiveness, which may explain why the hidden curriculum often remains hidden, it is also recognised that the hidden curriculum co-exists with the formal and informal curriculum (Elliot, 2022; Elliot et al., 2020). This also explains what underpins a possible scenario whereby two doctoral scholars can have a shared experience, e.g., participating in a workshop, but only one recognises and harnesses the workshop's implicit lessons. As a case in point, interaction with a workshop facilitator and participants may convey, even emphasise, the value of effective time management and impress on doctoral scholars how crucial it is. During the workshop, an implicit reference to the connection between a PhD and post-PhD life could stimulate further reflection. In turn, several workshop participants might come to appreciate that managing one's time is critical both for the doctoral journey itself and for post-PhD career planning. This then leads these doctoral scholars to manage their time actively and position themselves while preparing for a post-PhD career.

By envisioning what their post-PhD CV could look like, they strategise a doctoral journey that is aimed at completing doctoral research while embedding a plan to strengthen their subject knowledge and research skills and, in so doing, produce a tangible demonstration of knowledge and competencies (e.g., via publications, teaching experience) —characterising researcher competence. One may argue that enacting such reflection is informed by their agency and motivation (McAlpine & Amundsen, 2016). Equally, such profound reflection on both the doctoral journey and beyond is possibly stimulated by scholars' contemplation of the time management discussion during the workshop. Their participation and interaction with workshop participants served as a catalyst for such reflection. This is aligned to what Kuhn (2019) maintains, i.e., that critical thinking is 'a dialogic practice' where conversations with others enable a person to put forward their own argument (p. 146). This raises the question if there is a mindset, a disposition, a tacit knowledge, a skill, a personal quality, an inclination or a strategy that can help doctoral scholars to appreciate better the value embodied in an experience. Is this perhaps explained by the proverb: 'Beauty is in the eye of the beholder'? If so, what enables one person, but not another, to see such beauty?

A Theoretical Lens

In exploring this topic, I employ the concept of metacognition—a psychological construct I argue to be core not just to general effective learning but to achieving a major objective, such as managing one's entire doctoral experience efficiently (Elliot, 2022, 2023). According to the *Dictionary of Psychology*, in understanding metacognition, or the 'knowledge and beliefs about one's own cognitive processes', 'meta-memory' enables regulation of such cognitive functions in planning, checking or monitoring one's strategies (Coleman, 2015, p. 456). Building upon theory-of-mind development, this conceptualisation of metacognition goes back to John Flavell, who originally coined the term to denote active control of one's cognitive processes with a view to facilitating successful learning (Flavell, 2004). This definition suggests that metacognition entails having not just the ability but the intention to adopt and apply metacognitive skills. Likewise, it is worth noting that learners' capacity to regulate and employ metacognitive strategies is informed by their acquired knowledge through interaction with others, which influences their decisions, e.g., what to select and prioritise for future learning (Kirschner & Hendrick, 2020). McGahan and Stone (2022) further explain that metacognition refers to 'how learners can intentionally regulate their own cognitive skills to manipulate learning' (p. 177); in so doing, the emphasis is not only on learners' capacity to regulate their thinking but on it being done purposively to achieve an end. Very often, it is to advance learning. The anticipated added impact of applying metacognition makes understanding of this construct highly desirable. In the doctoral context that has knowledge production at its core, metacognition is arguably an indispensable tool for creating knowledge and advancing learning (Holmes et al., 2020).

At the same time, Kirschner and Hendrick (2020) stress the importance of understanding metacognition not just as a purely internal, cognitive process—learners' behavioural and social experiences need to be considered, too, as they are likely to have an impact on the process. As an example, the support that other stakeholders (e.g., supervisors and researcher developers) can offer in raising awareness of the necessity and implications of metacognition is essential. Such support forms a critical part of doctoral scholars' social experience—subsequently influencing both their thinking and behaviour. Put simply, understanding metacognition necessitates attention to the potential interaction between internal and external factors, i.e., one's cognition and other people's influence. In elucidating metacognition further, the argument I favour is that whereas acquisition of metacognitive skills and strategies, even competence, is good, having a disposition to employ metacognition is far superior (Kuhn, 2021). This, therefore, suggests that managing one's intentional usage of metacognition, rather than mere acquisition of metacognitive competence, can make a difference to scholars' doctoral experience. By 'making use of acquired control' or metacognitive strategies, new concepts, ideas and procedures can be harnessed (Kirschner & Hendrick, 2020, p. 247). This then suggests that metacognitive disposition 'puts scholars in a proactive mode as they consider and evaluate, plan, access and harness available resources to help address challenges encountered and find a resolution'. This idea is conveyed in the cyclical relationship involving 'appraise', 'solve problem' and 'revise'—emanating from scholars' metacognitive disposition' (Elliot, 2023, p. 159). Nevertheless, despite metacognition's emphasis on the internal processes to foster individual competence and autonomy, we need not underplay the idea that metacognition heavily relies on a person's exposure to other people's ideas and interaction with them. It then contributes to appreciation of what is valued and prioritised in certain contexts as well as ways of doing things, e.g., the standards to aim for, the goals they need to set for themselves, how to implement a chosen strategy and how to adjust their approach, if necessary.

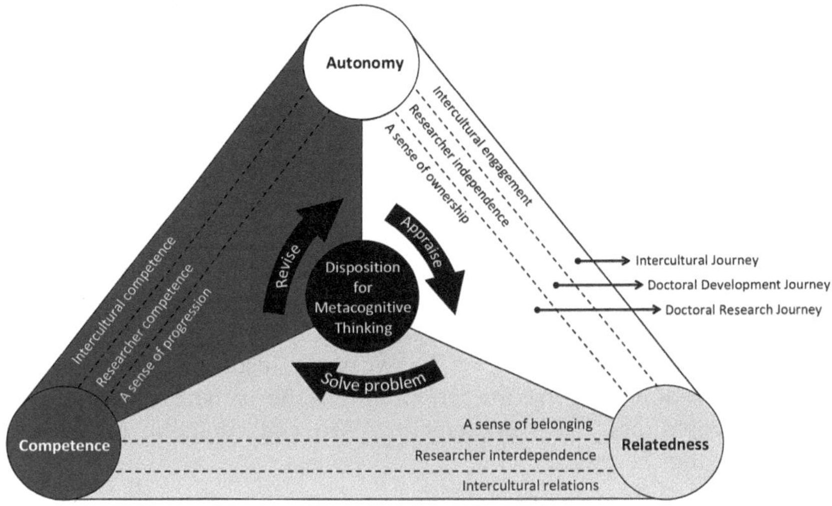

Fig. 1 A metacognitive approach to learning, motivation and intercultural relations: A new model for international doctoral scholars

Metacognition in Action

While Fig. 1 was originally employed to help elucidate how international scholars could navigate their doctoral experience, the metacognitive element—at the centre of the diagram—is arguably invaluable to all doctoral groups. More specifically, applying the iterative cycles of appraisal, problem-solving and revision can pave the way for doctoral scholars strengthening a sense of researcher independence. What this means in practice is that they consciously make an effort to reflect on their respective objectives in any of the three domains—doctoral research (its progression and completion), doctoral development (linked to post-PhD career preparation) and intercultural development (in the case of the international group). With these objectives in mind, scholars are then encouraged to adopt a cognitive habit, e.g., asking metacognitive questions in any situation they face:

- *Appraisal.* What personal strengths can I identify in myself given the current circumstances? Likewise, what are the areas in which I am lacking? Am I reliably appraising my strengths and goals based on my interaction with others? Where is development essential?

- *Problem-solving*. What are the ways in which I can build on my acquired personal strengths to respond more effectively to specific circumstances or challenges? How can I work with people around me to address the problem? What are the different possibilities available to me when seeking a resolution to a problem?
- *Revision*. Where did my attempt to resolve an issue lead to? Based on other people's successful cases, do I consider the strategy I adopted to be effective? If so, what can I learn from such a strategy? Can these lessons work in other contexts? On the other hand, if I regard my problem-solving efforts to be weak or insufficient, what other options do I need to consider? Are there alternative problem-solving strategies that I can implement? Who else could be involved in resolving the issues?

A Metacognitive Disposition as a Means of Harnessing the Hidden Curriculum

To put these exemplar metacognitive questions into context, let us consider a hypothetical case, i.e., a doctoral scholar's decision to apply for an internship with a view to acquiring knowledge, insights and skills that are not typically offered via institutionally offered courses and workshops. Megumi, a Year 2 doctoral scholar, has always wondered if applying as an intern for a journal editor would be either an advantage or a distraction that might keep her from concentrating on her doctoral work. Her response to the metacognitive cycle of questions guided her decision. In this case, the questions she asked and reflections she made, based on her discussion with her supervisor, included:

1. *Appraisal.* Knowing the importance of publication in doctoral and post-doctoral work, how much do I know about this process? Who are my potential sources of learning? By becoming an intern, am I likely to enhance my understanding of how publishing in a peer-reviewed journal works?
2. *Problem-solving*. Although my love of reading and writing led me to undertake a PhD, my knowledge of publishing and peer-review is almost non-existent. I reckon that pursuing an internship

opportunity, which entails constant interaction with a highly experienced author and exposure to the journal publishing process, will help me become familiar with this task that is often beyond the expectations for those pursuing a PhD monograph. The process also offers insights into what are considered acceptable standards in academic journals. Perhaps, my enhanced understanding can then increase my chance of getting published.

3. *Revision.* Upon reflection on my earlier experience as an intern to a journal editor, this led me to appreciate the rigour of the peer review process, e.g.,

    - various steps involved from initial assessment of the suitability of the manuscript (at times, leading to desk rejection, with reasons for the decision);
    - selection of anonymous reviewers based on subject and/or methods expertise;
    - several possible outcomes following the review—accept, minor revision, major revision, reject;
    - initial recommendation from editors (and co-editors) offering perspectives on their decision over the manuscript, at times, leading to seeking a new reviewer's perspective (in the case of a huge disparity among reviewers' decisions);
    - when authors are invited to address the comments from the reviewers, there remains the possibility that the manuscript will not be accepted if the reviewers and editors felt that the recommended changes were not adequately addressed.

These reflective questions led Megumi to evaluate and appreciate the overall value of engaging in journal internship and how it could enrich her doctoral research progress and her doctoral development as a scholar. Moreover, Megumi's internship further led to her valuing, enacting and pursuing unconventional, but invaluable, academic activities—particularly those that are not confined within the institutional context.

Following her internship, when Megumi returned to her initial 'appraisal' questions—

- Has my understanding of publishing in a peer-reviewed journal increased as a result of taking the internship opportunity?

- Has my experience helped not only in familiarising myself with a new academic experience but has it offered distinct insights that can increase my chances of getting published?

—Megumi's answers to these metacognitive questions are likely to inform her decisions concerning future opportunities. It could be another initiative that again differs from the courses and workshops offered by universities to doctoral scholars. This came from realising that participation in these initiatives offers an insight into activities pursued by, and typically becoming part of, the scholarly life of academic scholars, supervisors and other experts in the field. Upon further reflection, what Megumi did not expect is that her time as an intern in an academic journal also led to enhancing skills deemed invaluable by future employers including teamworking, working to a deadline, clarity of written expression, being systematic and organised.

Needless to say, her internship also expanded her network from whom she received informal advice when looking for the right journals or how to deal with unclear feedback from the reviewers or editors. Arguably, an expanded network opens more doors of opportunity—something discussed more comprehensively in Aarnikoivu's chapter. Not only did Megumi acquire concrete understanding from the journal internship itself but it also extended and strengthened her accumulated knowledge and repertoire of skills in journal publishing.

Taken together, Megumi's interaction with other scholars serves as resources that stimulated her metacognitive thinking and approach to learning advancement. Megumi's case then illustrates how having a disposition for metacognitive thinking can lead to genuine pedagogical lessons that can enrich one's doctoral learning experience. The benefits obtained from the journal internship were facilitated by a close and interdependent working relationship with the journal team. Significant learning started with scholars' openness to exploring new possibilities leading to cross-fertilisation of ideas.

En Route to Becoming Independent Researchers

In sum, within the doctoral context, harnessing these hidden lessons has been argued to complement or reinforce existing 'curriculum proper' or formal structures (Elliot et al., 2020; Martin, 1994). The challenge, however, is recognising, actively searching for and intentionally harnessing the hidden curriculum. General awareness of the contribution that hidden curricular lessons bring to doctoral scholars' development as competent and independent researchers is useful, but it is merely a starting point. Instead, it warrants a deeper appreciation of how supervisors (and other hidden curriculum agents) can encourage doctoral scholars to take advantage of genuine hidden curriculum opportunities. (In this connection, Albertyn elaborates on the idea of doctoral intelligence and their manifestations in her chapter.) Through supervisors' regular interaction with doctoral scholars, the question worth asking is—how can supervisors help instil the idea of intentionally harnessing the hidden curriculum as one of the significant pathways for becoming an independent and competent researcher?

Supervisors may proactively engage in more focused discussions specific to developing an active disposition for metacognitive thinking. Perhaps, regularly employing the metacognitive cycle of questions—appraisal, problem-solving and revision—as a guide for discussion is a way forward. These supervisor-initiated conversations can serve as ideal platforms for reflecting not only on the explicit benefits when attending workshops or taking part in internships or other opportunities (e.g., enhanced knowledge, research skills), but also in clarifying the impact of the implicit messages conveyed in these activities. Intentional metacognitive thinking can enable doctoral scholars to be more strategic in pursuing and harnessing hidden curricular lessons and, in turn, contributing to these scholars becoming more competent and independent researchers. In this respect, untapped resources, e.g., the supervisor's role (and that of other hidden curriculum agents) in cultivating a metacognitive disposition and realising this endeavour should not be underestimated.

Declarations of Conflicts of Interest Nothing to declare.

Acknowledgements Academic scholars—from the junior to the most experienced ones—typically learn from a combination of ways and resources. This book chapter has been inspired by the many doctoral scholars and colleagues, particularly Dr Kara Makara, with whom I have been fortunate to work closely for many years, enriching my scholarship and prompting me to reflect further. To you all, I am truly grateful.

References

Bengtsen, S., & McAlpine, L. (2022). A novel perspective on doctoral supervision. Interaction of time, academic work, institutional policies, and lifecourse. *Learning and Teaching in the Social Sciences (LATISS), 15*(1), 21–45. https://doi.org/10.3167/latiss.2022.150103

Coleman, A. M. (2015). *Oxford dictionary of psychology*. Oxford University Press.

Elliot, D. L. (2022). A 'doctoral compass': strategic reflection, self-assessment and recalibration for navigating the 'twin' doctoral journey. *Studies in Higher Education, 47*(8), 1652–1665. https://doi.org/10.1080/03075079.2021.1946033

Elliot, D. L. (2023). *Navigating your international doctoral experience (and beyond)*. Routledge.

Elliot, D. L., Baumfield, V., Reid, K., & Makara, K. A. (2016). Hidden treasure: Successful international doctoral students who found and harnessed the hidden curriculum. *Oxford Review of Education, 42*(6), 733–748. https://doi.org/10.1080/03054985.2016.1229664

Elliot, D. L., Bengtsen, S. S. E., Guccione, K., & Kobayashi, S. (2020). *The hidden curriculum in doctoral education*. Palgrave Pivot.

Flavell, J. H. (2004). Theory-of-mind development: Retrospect and prospect [Article]. *Merrill-Palmer Quarterly, 50*, 274+. https://link-gale-com.ezproxy.lib.gla.ac.uk/apps/doc/A121209520/AONE?u=glasuni&sid=bookmark-AONE&xid=70f39dd0

Holmes, P., Costa, N., & Lopes, B. (2020). The role of supervision in doctoral education: A transversal perspective. In M. Byram & M. Stoicheva (Eds.), *The doctorate as experience in Europe and beyond* (pp. 213–228). Routledge.

Kirschner, P. A., & Hendrick, C. (2020). *How learning happens: Seminal works in educational psychology and what they mean in practice*. Routledge.

Kuhn, D. (2019). Critical thinking as discourse. *Human Development*, 146–164. https://doi.org/10.1159/000500171

Kuhn, D. (2021). Metacognition matters in many ways. *Educational Psychologist*, 1–14. https://doi.org/10.1080/00461520.2021.1988603

Martin, J. R. (1994). *Changing the educational landscape: Philosophy women and curriculum*. Routledge.

McAlpine, L., & Amundsen, C. (2016). *Post-PhD career trajectories: Intentions, decision-making and life aspirations*. Palgrave Pivot.

McAlpine, L., & Amundsen, C. (2018). *Identity-trajectories of early career researcher: Unpacking the post-PhD experience*. Palgrave Macmillan.

McAlpine, L., & Norton, J. (2006). Reframing our approach to doctoral programs: A learning perspective. *Higher Education Research and Development, 25*(1), 3–17.
McGahan, J., & Stone, K. (2022). Cognitive models of learning. In J. Lord (Ed.), *Psychology of education: Theory, research and evidence-based practice* (pp. 161–184). Sage.

The Interstitial Doctoral Life of #thesisthinkers: When the Hidden Curriculum Might Be All There Is…

Tai Peseta, Giedre Kligyte, Amani Bell, Robyn Yucel, Gina Saliba, Fiona Salisbury, Brittany Vermeulen (nee Hardiman), Delyse Leadbeatter, Jenny Pizzica, Kate Thomson, and Sarah Barradell

#THESISTHINKERS: WHAT IS IT?

#thesisthinkers is the name we've given to our long-term scholarly collaboration together as doctoral scholars (or students), graduates, and supervisors in the field of HES. Currently, there are 11 of us: 4 supervisors, 4 PhD graduates, and 3 doctoral scholars, and we are all university workers. The #thesisthinkers' context is this: we're not all in the same

---

T. Peseta (✉) • F. Salisbury
Western Sydney University, Sydney, NSW, Australia

The University of Sydney, Sydney, NSW, Australia
e-mail: t.peseta@westernsydney.edu.au; F.Salisbury@westernsydney.edu.au

D. L. Elliot et al. (eds.), *Developing Researcher Independence Through the Hidden Curriculum*,
https://doi.org/10.1007/978-3-031-42875-3_10

department (or an academic one at all in some cases) or even in the same University (we work and study across 6 universities at last count). Many of us have uneven attachments to HES as our primary field—especially where there are cognate fields and professions that have shaped our subjectivities (including design, science, secondary education, philosophy, library studies, and allied health among them). For some of us, #thesisthinkers is the only research community that supports our labour, identities, and projects as doctoral supervisors and scholars. It is not sanctioned as a supervision pedagogy by any Department, School, or Faculty; it doesn't count as a developmental activity on any institutional register for supervision development; there's no milestone requirement that mandates student participation in it as part of their candidature, and the supervisors involved do

G. Kligyte
The University of Sydney, Sydney, NSW, Australia

University of Technology Sydney, Ultimo, NSW, Australia
e-mail: giedre.kligyte@uts.edu.au

A. Bell • D. Leadbeatter • K. Thomson
The University of Sydney, Sydney, NSW, Australia
e-mail: amani.bell@sydney.edu.au; delyse.leadbeatter@sydney.edu.au; kate.thomson@sydney.edu.au
R. Yucel
Deakin University, Melbourne, VIC, Australia
e-mail: robyn.yucel@deakin.edu.au

G. Saliba
Western Sydney University, Sydney, NSW, Australia

Australian College of Applied Professions, Sydney, NSW, Australia
e-mail: Gina.Saliba@acap.edu.au

B. Vermeulen (nee Hardiman)
Western Sydney University, Sydney, NSW, Australia
e-mail: B.Vermeulen@westernsydney.edu.au

J. Pizzica
Sydney, NSW, Australia

S. Barradell
Swinburne University, Melbourne, VIC, Australia
e-mail: sbarradell@swin.edu.au

not even receive workload for the work that goes into establishing, nurturing, and maintaining #thesisthinkers. One might say that the work of #thesisthinkers is hidden, underground, and unacknowledged. Yet between 2016 and 2022 through sheer persistence, #thesisthinkers will have acted as the main pedagogical mechanism to support 4 part-time doctoral scholars to complete, and together and separately, we will have produced over 85 peer reviewed scholarly outputs. While completion and publication are often standard metrics for research success, much more interesting to us is how a pedagogical practice like #thesisthinkers—one that has an interstitial life in the curriculum underground—survives and remains productive as a form of interdependence. Given its institutional slipperiness, what might explain how #thesisthinkers continues to sustain itself?

What Do We Do Together in #thesisthinkers?

First, it is important to note that #thesisthinkers complements the arrangements of individual doctoral supervisors and scholars (and those in panel mode too). It is not a substitute for the labour that is specific to the close-up work generated by supervision's attention to the scholar and their production of text (or designed to be one); rather, #thesisthinkers is intended to articulate, join, extend, and challenge those conversations—it is a place of communal puzzling infused with the generosity and interdependence of co-inquiry. Second, as much as #thesisthinkers supports doctoral scholars' candidature and progress, it can also be understood as a place of shared supervision pedagogy—where supervisors participate in learning about supervision alongside the company of colleagues. Third, what we also have in common is a curiosity about the field of HES, and how our projects enact care and challenge for a field of inquiry (Peseta et al., 2021). #thesisthinkers brings together scholars and supervisors who are exploring the way HES may be the field they are producing new knowledge in, for, and about. (See also Frick's chapter discussing creative pedagogues and Wisker's chapter on 'Fridaying'.) That commitment and context can change as the research proceeds and new paths are explored and forged. So, what do we do together? Is it all that unusual?

Since 2016, we have met for 2 hours each month, sharing the responsibility for curating a collective agenda. The first hour is typically devoted to one or two doctoral scholars' work—a draft article, thesis chapter, conference paper, a discussion about data, methodology, conceptual and theoretical spines—something a doctoral scholar has written (or is thinking about) and is seeking conversation to take the next step. Their piece is circulated a week before the meeting, alongside a set of questions/puzzles they are inviting us into. The second hour is more variable. Over the years, we have engaged with research topics and doctoral practices that run across our projects (e.g., reflexivity, identity, theory, publishing, reviewing, impact) as a block (say for 3 months) or as something more one off. As we puzzle our way through the particularities of being HE researchers and how our projects contribute to the field, we also engage with the scholarly literature about doctoral life and with practitioners who offer insight about it.

Since 2017, we also make time for a dedicated four days away writing together, twice a year. Based on Grant's (2006) Women Writing Away residential model, we commit to a set of writing goals before we go on retreat so that our energies are focused on writing. Structured mainly around individual writing time, with optional and short activities (for new researchers, i.e., stylish writing; first sentences), a Critical Friend pairing for the entire retreat, a Slow Reading session over wine and chocolate (we read an article together, each person reads a paragraph out loud, after each one we pause to discuss it before moving to the next), a compulsory work-in-progress session (in a trio, each participant has 45 minutes to bring a puzzle from their writing for collective problem-solving and reflection), a trivia night, and much laughing and chat over food and wine, our aim is to reconnect with our writing lives as researchers without the daily grind of distraction. We do not count word production like some doctoral retreats and writing boot camps do (we can see some merit in the pleasure of watching a daily word count increase). Instead, we design the retreats in ways that move us between the solitude a writer so often desires, the conviviality we crave, and the responsibility of accounting for our writing and thinking to each other while on retreat.

What Makes #thesisthinkers Sticky?

There are likely five things.

The Work of Usefulness Is Shared and Co-created, and It Shifts

#thesisthinkers keeps going because it does something useful. Leaning on Roxå and Mårtensson's (2015) work on microcultures in HE where aspects of trust, significance, and shared responsibility for a collective developmental agenda are crucial, we too must decide our 'usefulness' as a way of showing how, and why, we matter to one another. Each of us has a hand in keeping #thesisthinkers going, making it work, and contributing our labour to it. No doubt there have been moments of frustration, of waning energy, and of asking: Should we keep this going? Of what 'use' is this to me, and to us, now? On 'use', Ahmed (2018) writes:

queer use [is] the work you have to do to be. The more you are blocked, the more you have to try to find a way through. The less support you have the more support you need. We might become each other's resources, we prop each other up, because we understand how diminishing it can be to have fight for an existence, to have to fight, even, to enter a room.

For some of us, the act of keeping #thesisthinkers alive and active is our way of remembering, declaring, and keeping alive our identities, practices and contributions as researchers and supervisors in work environments that can be ambivalent at best, and hostile at worst.

It's Specific to Individuals' Research Projects

To complement individual supervision, many universities (including the ones we work in) offer an array of centralised support and conversation for both doctoral scholars and supervisors. Workshops, mentoring, modules, seminars, Shut Up and Write sessions, conferences, and the like now litter the online international environment. The response to COVID has also made it possible for doctoral scholars to access a larger community of researchers. In #thesisthinkers, we aim to keep the focus (and the expectation) on care for the student, their research project, and their wrestling with research as it plays out alongside the demands of new knowledge production. The time in #thesisthinkers monthly meetings is for doctoral scholars to exercise their judgement while the rest of us are encouraged and facilitated to see how that student's wrestling offers lessons for our own research.

It's Relational

The PhD is a long commitment, and we need colleagues beside us. Over the course of #thesisthinkers, we have gotten to know one another. We recognise Giedre's theoretical writing demands a certain kind of clear-headed attention from us. We know Fiona has just finished a writing course with a noted Australian poet and we notice the transformations to her writing. We are alongside Delyse as she shapes a post-PhD research programme. We share in Robyn's joy at wonderful examiners' reports. We are there as Britt and Gina encounter the frustration and minutiae of an ethics application, and we marvel at how disciplined Sarah is in publishing articles to be included in her thesis. In summary, we commiserate and celebrate when one of us reaches a milestone or achievement: graduation, a new job, a successful grant, an academic promotion, a book, article, and on occasion, a new family addition. While #thesisthinkers draws us together as HE researchers and workers, the conversations consolidate into a connection with each other that travel well beyond the PhD.

Outputs Are Not the Focus: Cultivating Good Thinking and Academic Judgement Are

While we are proud of the 85+ scholarly outputs we have generated, at the heart of #thesisthinkers is feeding our research imaginations so that our contributions prise open an inquiry for others. It has become too easy for doctoral scholars to be inducted into a research disposition that pursues quantity and outputs that is now endemic in an academic culture of over-work and over-production. Our focus together in #thesisthinkers is to develop our judgements about what good thinking is, to practise justifying our judgements to one another as researchers, and to learn from the responses offered by others. In the main, the outputs follow the thinking; although on occasion, they can also function to push along the communication of our thinking too.

It's Expansive

At one level, the endpoint of #thesisthinkers for doctoral scholars (and their supervisors) is graduation. It marks the end of the doctoral road and launches the beginning of a life as an independent researcher. Yet, that is not the only narrative we cultivate in #thesisthinkers. We encourage colleagues to continue and reset their participation as an opportunity to craft the next phase of their research life. This might include shaping a new programme of research, taking a step into a funded grant collaboration, or indeed, becoming doctoral supervisors themselves and bringing their own students to #thesisthinkers as an act of expansion. The act of continually co-creating a collective #thesisthinkers agenda suggests that these research transitions can be accommodated (side note: all of us currently work at a university, and we've not encountered yet a non-university post-PhD life in the way doctoral scholars in other fields may be likely to).

#THESISTHINKERS: A CURRICULUM HIDING IN PLAIN SIGHT?

In our deliberations about #thesisthinkers as a productive interstitial doctoral space, it is worthwhile reflecting on the conditions and boundaries that mark some aspects of learning (and curriculum) as secret and visible, intended and unintended, formal and informal. In much of the scholarly work about the hidden curriculum in doctoral education, there is very often a distinction about the location of students' learning experiences (in or outside of the academy), and whether the institutional structures and systems set up for doctoral scholars' learning, signals an intention by the university for students to learn something in particular (usually related to academic knowledge). Below, Eliot, Bengsten, Guccione, and Kobayashi (2020) offer their definition:

> we regard the hidden curriculum in doctoral education as the unofficial (and informal) channels of genuine and useful learning that can be acquired within or outwith both the physical and metaphorical walls of academia. By contrast, the formal curriculum from which knowledge is specifically gained by study refers to activities where learning is typically acquired via the official (or structured) doctoral courses, seminars, workshops, and supervisory meetings—strictly within what we regard to be the academic setting. (p. 4)

While Elliot et al.'s (2020) distinction is no doubt analytically helpful, by this definition, #thesisthinkers effectively confounds these boundaries. They make far more sense when the supervision relationship appears more typical: where supervisors and students are co-located in the same university, and together, they can take advantage of the structures and resources on offer as part of an institution's doctoral programme. In #thesisthinkers' case, the supervisors often work at a different university than where the student is enrolled; in some cases, there is no local discipline-based expertise where the student is enrolled; and owing to students' part-time candidature, their access to, and subsequent participation in, a research culture is limited. #thesisthinkers is a response to precisely these kinds of anomalies.

Because of our context, there is great deal about #thesisthinkers that is intentional, but we are invited to adapt to curriculum systems that do not acknowledge our cross-institutional circumstances or its particularity. #thesisthinkers emerged exactly because the doctoral education and supervision pedagogy we wanted was absent from the menu of experiences our institutions had laid out for us. Indeed, we had to co-create the research relationships, interdependencies, and cross-institutional conversations we were seeking. Despite its lack of institutional recognition, #thesisthinkers is sustained by us because it is local, emergent, co-created, useful, and the care for it, is shared among trusted friends and colleagues. To our minds, it is exactly the kind of curriculum that a doctoral education invites.

References

Ahmed, S. (2018, October 18). *What's the use? Feminist killjoys.* https://feministkilljoys.com/2018/10/18/whats-the-use/

Eliot, D., Bengsten, S. E., Guccione, K., & Kobayashi, S. (2020). *The hidden curriculum in doctoral education.* Palgrave Macmillan.

Grant, B. M. (2006). Writing in the company of other women: Exceeding the boundaries. *Studies in Higher Education, 31*(4), 483–495. https://doi.org/10.1080/03075070600800624

Peseta, T., Kligyte, G., Bell, A., Hardiman, B., Leadbeatter, D., Pizzica, J., Saliba, G., Salisbury, F., Thomson, K., & Yucel, R. (2021). Borders, paths and orientations: Assembling the higher education research field as doctoral students and supervisors. *Teaching in Higher Education, 26*(3), 422–437. https://doi.org/10.1080/13562517.2021.1885022

Roxå, T., & Mårtennson, K. (2015). Microcultures and informal learning: A heuristic guiding analysis of conditions for informal learning in local higher education workplaces. *International Journal for Academic Development, 20*(2), 193–205. https://doi.org/10.1080/1360144X.2015.1029929

PART III

Insights from Researcher Developers

The Dance of Authenticity and Multiple Ways of Doing: Defining a Pedagogy for Accessing the Hidden Curriculum in Doctoral Education and Developing Researcher Independence

Kelly Louise Preece

This chapter will explore how I developed and defined a pedagogy for the Researcher Development Programme (RDP) at the University of Exeter. The RDP 'is designed to support [postgraduate researchers] at all stages of [their] research degree' (University of Exeter, n.d.). I led the RDP from 2015 to 2022, during which I redesigned the programme so that it enabled doctoral scholars to bridge the conceptual boundary between formal and informal learning by introducing multiple 'ways of doing' and thereby uncovering the Hidden Curriculum. This chapter will trace the pedagogical roots of this approach back to my disciplinary upbringing in

K. L. Preece (✉)
University of Exeter, Exeter, UK
e-mail: k.preece@exeter.ac.uk

D. L. Elliot et al. (eds.), *Developing Researcher Independence Through the Hidden Curriculum*,
https://doi.org/10.1007/978-3-031-42875-3_11

contemporary dance, specifically improvisation. I will argue that by introducing researchers to multiple ways of doing, combined with the practice of authentic teaching, we empower them to find ways of working that suit their research context and their preferences and, in doing so, develop independence, here conceptualised as taking responsibility for their research and their learning, and through this becoming confident to improvise from a place of authenticity in their professional lives.

'The Only Way to Do It Is to Do It' (Merce Cunningham)

When I started my role as Researcher Development Manager at the University of Exeter, I inherited a (prolific) collection of PowerPoint slides for training sessions on everything from conducting a literature review to writing an academic CV. Despite the range and scope of these materials, I was struck by the vagueness inherent within them. This is in no way to criticise my colleagues: on the contrary, as I immersed myself in the resources I had inherited, I found so much that was useful, insightful, and important for doctoral scholars. All the materials that I inherited emphasised the need for researchers to take responsibility for their own learning, and to critically reflect on their approaches and practices. The programme, like many in the UK, is underpinned by the Vitae Researcher Development Framework (RDF) which 'describes the knowledge, behaviour and attributes of successful researchers' (Vitae, 2023). The RDF is used nationally to develop policy and practice, as well as being a tool for researchers to critically reflect on their strengths and areas for development. In focusing on this process of critical reflection, I provide opportunities for researchers to discuss and reflect on the different possible 'ways of doing', offering options with which doctoral scholars could experiment and choose an approach that was relevant to their disciplinary context, and to their working and learning styles. This approach to reflection, grounded in practice and ways of doing, is not new to education scholars—indeed, the concept of experiential learning (Kolb, 1984) is well trodden throughout educational literature. But it is not from the experiential learning cycle that I developed the pedagogy of the Researcher Development Programme. Rather, it is my background in contemporary dance, where the knowledge and practice of embodied and experiential learning and knowing by far pre-dates Kolb's writing.

I began my career as a Lecturer in Dance, having worked since my early teens as a professional dancer and later as a choreographer. Teaching and learning in contemporary dance are rooted in 'embodied practice' in the studio (Bannon, 2010, p. 49), as well as reflection-in-action and reflection-on-action in the tradition of Donald Schon (1983). This reflexivity is central to contemporary dance and draws on a variety of schools of thought about learning and knowledge-making as an *experiential* process, including embodied knowing (Davidson, 2004 and Parviainen, 2002) and practice-as-research (Nelson, 2009). As a student and a teacher my practice was and is improvisation—the practice of 'composing whilst dancing' (Buckwalter, 2010). Improvisation is plagued by misconceptions that its practitioners engage in nothing more than 'aimless, even talentless, noodling' (Gere, 2003, p. xv). However, improvisation practitioners know that improvisation is all about rules or 'predetermined overarching structural guidelines that delimit the improvising body's choices, such as a score for the performance, or any set of rules determined in advance' (Foster, 2003, p. 4). In other words, improvisers know that we make our most interesting and creative choices when experimenting within a set of constraints. A task I returned to in every class was called the Slow Journey, developed from the book *Body Space Image: Notes Towards Improvisation and Performance* (2014) by Miranda Tufnell and Chris Crickmay. Near the start of every class, I would task students with the simple task of moving between standing and lying on the floor—but doing so as slowly as possible over 5 minutes.

Framing Improvisation

Consider that if I were to ask you now to stand up and improvise something, you would likely be unsure where to begin. When faced with unlimited options and possibilities, we are blinded by choice and often paralysed and unable to act. However, if I were to ask you to move to the other side of the room without your feet touching the floor, now you have a task, a goal, and a framework something concrete upon which to act. And even if every reader of this book were all given the same set up and the same room, it is unlikely that any two would approach the task in precisely the same way.

Improvisation as taught in contemporary dance requires critical reflection in action, to enable its practitioners to make movement choices, choices informed by years of training and experience. One of my first academic colleagues, Professor Fiona Bannon, once said the same to me of research. In one of many influential conversations in her office at the University of Leeds, Fiona proposed that she could start a group of doctoral scholars off on the same research project, and everyone would produce a different thesis because they would have different experiences and training to bring to the topic and would therefore make different choices through the research process. Just as a dancer has their training and technique upon which to draw, researchers need a concrete grounding upon which to reflect and act. While some of the tasks and skills required of postgraduate researchers may be familiar, so many are not; they are part of the Hidden Curriculum. For example, networking at conferences is an important part of developing your research profile, and most institutions offer workshops on networking or 'Making the most of conferences'. These workshops offer multiple ways of engaging—including introducing yourself to another scholar, asking your supervisor to introduce you to their colleagues, or using social media to develop relationships and networks within the academic community. Different approaches are more comfortable, and more appropriate, for different researchers, different disciplines, and different contexts. Like improvisers, researchers need something concrete and practical upon which to reflect and make choices—here, about how to approach networking. This was the basis upon which I developed a pedagogy for the RDP, which accessed the Hidden Curriculum and ultimately developed researcher independence, through framing the options, the importance of choice, and the influence of the self, in making research and career decisions.

I argue below that this need for choice and personalisation is as true for the pedagogical choices we make in developing researchers and research skills as it is for the process of research.

A Pedagogy of Multiple Ways of Doing

In supporting our postgraduate researchers at Exeter, particularly those just starting out on their doctoral journey, I wanted to elucidate the Hidden Curriculum for them whilst moving through the four stages of competence cited in Elliot et al., 2020, pp. 10–11—in particular passing the threshold between stage 2, conscious incompetence ('I'm attending this literature review course as I don't know where to begin'), and stage 3, conscious competence ('I am confident in my approach to searching and selecting literature) (Castle & Buckler, 2018, p. 54; Clarkson & Gilbert,

1988 cited in Donati & Watts, 2005, p. 478). I argue that to effectively pass through this threshold, doctoral scholars need something concrete upon which to base their reflection *on*; the equivalent of an improviser's rules, structure, or score. Many doctoral scholars will come to doctoral study without having undertaken a literature review—or if they have completed one before, it is unlikely they will have been asked to construct one around a gap in knowledge, given that the requirement for originality is distinct in doctoral study. Achieving the transition to conscious competence can be supported through offering a range of opportunities for personalisation, for example, for how to take and organise notes for a literature review, or by helping researcher to build their own strategies to identify thesis structure. I call this introducing doctoral scholars to 'multiple ways of doing'.

You could argue that this 'multiple ways of doing' sounds a lot like the formal, structured learning or curriculum of seminars, workshops, and supervisory meetings rather than the Hidden Curriculum. I would, however, assert that this approach crosses the conceptual boundary splitting the two (Elliot et al., 2020, p. 7). The formal curriculum guides researchers towards the product—for example, to complete a doctorate, a researcher must produce the 'product' of a literature review. The Hidden Curriculum represents 'the how', offering students different ways of doing—of searching literature, note-taking, or structuring their writing that guide them through the process (Elliot et al., 2020, p. 8). The formal curriculum is the task; the informal curriculum is the multiple 'ways of doing' that bring researchers to that product or outcome. Through this, multiple ways of doing becomes a framework for building researcher independence, through considered choice. By providing multiple ways, rather than one way, to access the Hidden Curriculum, and giving researchers examples of the tools they might use to get there, they are given something concrete upon which to reflect and make choices—therefore taking responsibility for their choices in research and in their professional learning and thereby developing independence. (These principles are in line with adopting the four domains of doctoral intelligence discussed in Albertyn's chapter.)

Example: Introducing Strategies to Identify Thesis Structure

When I have run workshops on 'writing your thesis', one of the main concerns doctoral scholars arrive with is how to structure their thesis. Whilst there are disciplinary norms, there isn't 'one way' to structure a doctoral thesis—just like there isn't 'one way' to do a literature review. The thesis structure needs to be responsive to the research, and so doctoral scholars need to be given the tools to critically assess their research material to develop a thesis structure that most effectively tells the story of their research—and their original contribution. I do this by:

- Introducing the institutional regulations and basic structures
- Introducing a collation of ways to identify a structure for your thesis including:
  - Starting from argument and contribution—with examples from Susan Carter (2018) and Inger Mewburn (2016)
  - Analysing existing theses—with examples from Cally Guerin (2018) and Vitae (n.d.)
  - Mind mapping your thesis—with examples adapted from Pat Thomson (2016)
  - Research storyboarding—with examples from Patrick Dunleavy (2017)
  - Thesis skeleton—with examples from Steve Draper (2003)
- Offering a case study/example from a doctoral graduate
- Providing a blank template and activity outlines for trying and evaluating the different strategies

This example provides a grounding upon which for doctoral scholars to reflect through different tools they can use to identify their structure—ways of doing—alongside an example of this in practice. In doing so, it takes postgraduate researchers from conscious incompetence (I have no idea how to structure my thesis) to conscious competence ('I have a range of tools I can use to identify my thesis structure, that I can use and adapt to suit my research and my preferences'). You can read a full version of this example developed into an online resource (Preece, 2022). As exemplified here, the author also served as a creative pedagogue of the Hidden Curriculum; see Frick's chapter.

An Authentic Approach to Training and Development

Explicit engagement with the Hidden Curriculum, then, is a vital part of how I developed a programme that fostered independence in decision making. Alongside the pedagogy of 'multiple ways of doing', my own manner in the classroom is crucial to my pedagogy and how I create effective learning environments for accessing the Hidden Curriculum and developing independence. This is because my teaching 'style', grounded in authenticity, represents my own choices, carefully selected from multiple ways of doing—of teaching. In their research on teacher authenticity, Johnson and LaBelle define authenticity in the classroom as 'existing in a way that is consistent with one's own thoughts, feelings, emotional and overall sense of self' (2017, p. 424), this is turn creating 'a more open and supportive classroom' (2017, p. 426). They identify five sets of behaviours associated with authentic teaching: approachability, passion, attentiveness, capability, and knowledge (2017, p. 429). Through sharing personal stories of academic success and failure, being uninhibited in my enthusiasm for my work, willingness to share my knowledge and expertise, being open to 1-2-1 conversations and a gentle use of humour (see Kobayashi and Berge's chapter on humour supporting researcher independence), my Researcher Development workshops are a safe space for students to share their own experiences and receive advice, support and ultimately problem-solve. As a practitioner of authentic teaching, I consider this approach crucial to facilitating the pedagogy of 'multiple ways of doing'. Being present in the moment, responding to situations arising, and carefully choosing my response is a form of improvisation. This approach creates an open environment for dialogue, reflection, and sharing. (See also Makara et al.'s chapter for another example of using authentic and safe space for doctoral learning.) After a workshop on presentation skills, one postgraduate researcher commented on the feedback form that the session was:

> [g]enuinely inspiring. It was great to be told that it's ok to be yourself onstage and that you don't have to be a spotlight-hogging extrovert to give a good presentation—it felt like if you were naturally a thoughtful quiet person, or sarcastic, or had a sense of humour, or were passionate about a

> particular aspect of a topic, it would be ok to let that come across appropriately in your presentation and not just conform to 'what you think a lecturer MUST sound like'.

Within my authentic approach, the postgraduate researcher was able to identify multiple ways of doing, and use my authenticity, the embodiment of my own choices, to make informed decisions about their own approach to delivering presentations. In doing so they took responsibility for their research and learning, and therefore moved towards independence.

Conclusion

In this chapter I have articulated a pedagogy to develop researcher independence by offering up multiple ways of doing. I am reminded of choreographer Twyla Tharp's proposal that '[c]reativity Is a habit, and the best creativity is a result of good work habits' (Tharp, 2006, p. 7). Tharp's assertion that creativity is a habit seems paradoxical, and '[w]e think of creativity as a way of keeping everything fresh and new, whilst habit implies routine and repetition' (Tharp, 2006, p. 9). Tharp elucidates how creativity is not necessarily a bolt of lightning from above, but is instead a product of habit and routine. Through habit and routine, we make the conditions for creativity to emerge—in the same way that in improvisation we create the conditions for new movement patterns through rules, structures, and scores. In fact, Tharp's approach to and seminal book on creativity is something I refer to regularly when discussing project and time management with doctoral scholars. I argue that in developing habits and routines for their working day, they are setting themselves up to be creative and productive in their work, by making informed choices from a range of options and being aware of their own authentic style; this empowers them to confidently improvise in their research lives. (See also Frick's chapter on creative supervising and supervising for creativity.)

To return to the Hidden Curriculum, through offering multiple ways of doing, we create the conditions for researchers to develop their independence as scholars. We give them options to enable them to make choices.

Acknowledgements My thanks to Dr Caitlin Kight and Dr Edward Mills for their feedback in the development of this chapter.

But most importantly, my deepest thank you to all the doctoral scholars I have had the privilege of working with throughout my time leading the Researcher Development Programme at the University of Exeter.

References

Bannon, F. (2010). Dance: The possibilities of a discipline. *Research in Dance Education, 11*(1), 49–50.

Buckwalter, M. (2010). *Composing while dancing: An improviser's companion*. University of Wisconsin Press.

Carter, S. (2018). Turning facts into a doctoral story: The essence of a good doctorate. *Doctoral Writing SIG*. https://doctoralwriting.wordpress.com/2018/02/19/turning-facts-into-a-doctoral-story-the-essence-of-a-good-doctorate/

Castle, P., & Buckler, S. (2018). *Psychology for teachers* (2nd ed.). Sage.

Davidson, J. (2004). Embodied knowledge: Possibilities and constraints in arts education and curriculum. In L. Bresler (Ed.), *Knowing bodies, moving minds* (pp. 197–212). Kluwer Academic Publishers.

Donati, M., & Watts, M. (2005). Personal development in counsellor training: Towards a clarification of inter-related concepts. *British Journal of Guidance & Counselling, 33*(4), 475–484.

Draper, S. (2003). *Creating the skeleton for a thesis*. Website. https://www.psy.gla.ac.uk/~steve/resources/skel.html

Dunleavy, P. (2017). Storyboarding research. *Writing for Research*. https://blogs.lse.ac.uk/writingforresearch/2017/07/17/storyboarding-research/

Elliot, D., Bengtsen, S., Guccione, K., & Kobayashi, S. (2020). *The hidden curriculum in doctoral education*. Palgrave Macmillan.

Foster, S. (2003). Taken by surprise: Improvisation in dance and mind. In A. Cooper Albright & D. Gere (Eds.), *Taken by surprise: A dance improvisation reader* (pp. 3–12). Wesleyan University Press.

Gere, D. (2003). Introduction. In A. Cooper Albright & D. Gere (Eds.), *Taken by surprise: A dance improvisation reader* (pp. xiii–xxi). Wesleyan University Press.

Guerin, C. (2018). Reading theses to write a thesis. *Doctoral Writing SIG*. https://doctoralwriting.wordpress.com/2018/05/14/reading-theses-to-write-a-thesis/

Johnson, Z. D., & LaBelle, S. (2017). An examination of teacher authenticity in the college classroom. *Communication Education, 66*(4), 423–439.

Kolb, D. (1984). *Experiential learning*. Prentice-Hall.

Mewburn, I. (2016). Better than Donald—How to argue like a pro. *The Thesis Whisperer*. https://thesiswhisperer.com/2016/11/09/better-than-donald-how-to-argue-like-a-pro/

Nelson, R. (2009). Practice-as-research knowledge and their place in the academy. In R. Nelson (Ed.), *Practice-as-research: In performance and screen* (pp. 112–130). Palgrave Macmillan.

Parviainen, J. (2002). Bodily knowledge: Epistemological reflections on dance. *Dance Research Journal., 34*(1), 11–26.

Preece, K. L. (2022). *Structuring your thesis.* Website. https://researcher-development.co.uk/writingyourthesis/structuring-your-thesis/

Schon, D. (1983). *The reflective practitioner.* Basic Books.

Tharp, T. (2006). *The creative habit.* Simon and Schuster.

Thomson, P. (2016). Mapping your literatures. *Patter.* https://patthomson.net/2016/04/04/mapping-your-literatures/

Tufnell, M., & Crickmay, C. (2014). *Body space image: Notes towards improvisation and performance.* Dance Books Ltd..

University of Exeter. (n.d.). *Training and development.* Website. https://www.exeter.ac.uk/research/doctoralcollege/researcherdevelopment/training/

Vitae. (2023). *About the researcher development framework.* Website. https://www.vitae.ac.uk/researchers-professional-development/about-the-vitae-researcher-development-framework

Vitae. (n.d.). *Getting starting on writing your thesis.* Website. https://www.vitae.ac.uk/doing-research/doing-a-doctorate/completing-your-doctorate/writing-and-submitting-your-doctoral-thesis/getting-started#analysing-a-thesis

How Humour Can Support Researcher Independence

Sofie Kobayashi and Maria Berge

Humour and Norms in Supervision

Doctoral scholars need to know the norms and standards of their discipline to become independent researchers. Norms are arguably part of the hidden curriculum in doctoral education, as also noted by Anderson when she describes norms in academia: 'Neither a normative system nor an individual's normative orientation is fully knowable since many of a social system's norms remain latent until they are challenged or violated' (Anderson et al., 2010, p. 367). Norms of the discipline are closely linked with standards (Kobayashi & Emmeche, 2023). Gurr (2001) has

S. Kobayashi (✉)
Department of Science Education, University of Copenhagen, Copenhagen, Denmark
e-mail: skobayashi@ind.ku.dk

M. Berge
Department of Science and Mathematics Education, Umeå University, Umeå, Sweden
e-mail: maria.berge@umu.se

D. L. Elliot et al. (eds.), *Developing Researcher Independence Through the Hidden Curriculum*,
https://doi.org/10.1007/978-3-031-42875-3_12

proposed a 'supervisor-student alignment model' with the objective of improving supervisory practice and supporting the development of doctoral scholars' independence. In his research 'competent autonomous' means that doctoral scholars are 'cognisant of the norms, expectations and standards within their discipline and are able to assess their own plans and actions to ensure compliance with these' (p. 85).

An understanding of norms and expectations is not only desirable—but crucial knowledge for building independence as a researcher (Gurr, 2001). In this chapter we show how humour and laughter can facilitate development of doctoral scholars' independence. Humour is an aspect of interpersonal interaction that we often use unconsciously, but humour and laughter play an important role in interaction. We argue that humour has an important connection to norms and values when learning together, since humour is a way of demonstrating understanding within a group, but also negotiations of boundaries (Billig, 2005).

An important function of all humour is to create a sense of belonging when laughing together (Plester, 2015). Humour can also support the learning processes in supervision, as we see in our study of doctoral scholars learning about norms of research during supervision (Kobayashi & Berge, 2022). In this study, one of our informants told us that he as a supervisor explicitly tells all his doctoral scholars that he hopes for good humour, informal atmosphere, and to be open for mutual teasing as part of his effort to align expectations between him and his doctoral scholars at the beginning of the supervisory process. Li and Seale (2007) have also illustrated how laughter and humour have an important function in the interaction when supervisors are delivering criticism without 'threatening social solidarity' (p. 520). Humour is about entering a grey zone defined by norms in that specific context (Billig, 2005), and laughter indicates that such a line has been surpassed (Berge, 2017).

Norms are an aspect of interaction that do not receive much attention either, although compliance with the norms of science is crucial for good research and for becoming an independent researcher (Gurr, 2001). Studies in the sociology of science lay out the general norms of science that scientists strive for: that scientists share new knowledge, that research should be evaluated on its own merits, that scientists are motivated by discovery rather than the possibility for personal gain, and that scientists consider all new evidence, hypotheses, theories, and innovations, even those that challenge or contradict their own work (Merton, 1942; Anderson et al., 2010).

These norms all have counternorms that researchers may subscribe to in some situations. More norms have been amended, for example, 'originality' (Merton, 1957), 'calling' (the passion for science) versus 'employment' (a job to be finished), and 'quality' (good publications) versus 'quantity' (salami slicing research for more publications) (Anderson et al., 2010). In this chapter we look at the disciplining effect of humour, where humour highlights important aspects of grey zones in the interaction, such as norms and standards in research. (See Preece's chapter for a brief discussion of how gentle humour is typically employed as a personal strategy for building an authentic and safe teaching environment.)

Examples from Life Science Supervision

In what follows here we give examples of how humour can be part of building independence in doctoral supervision. The examples are all taken from four recorded observations of doctoral supervision in life science (named Cases 1–4). Our methodological starting point was that laughter *transforms* what had been said, into humour in the interaction (Berge, 2017), and consequently we have looked for laugh units (Jefferson et al., 1977), from giggles to guffaws, in all four cases.[1] Hence, the norms, expectations, and standards become visible to the persons in the conversation through laughter—and to us as researchers analysing the interaction. Here we focus on illustrating how humour can support independence. When presenting our examples of humour in supervision we are using acronyms for the doctoral scholars (DS) in the four cases: Anne (Case 1), Aisha (Case 2), Sara and Anish (Case 3), and Cecilie, Hans, and Postdoc Kristian (Case 4). We have kept the titles of supervisors and co-supervisors to indicate relations.

In our first example of humour doctoral scholar Aisha suggests her supervisors look at her material with a giggle:

DS Aisha:	would you like to see my clinoptilolite?
Co-supervisor:	oh
DS Aisha:	**HeHe**, it's here
Co-supervisor:	oh jah? this is the one you used in [country]?

Case 2

[1] We are aware that this method, to recognise laugh units rather than to code what is objectively 'funny', has the limitation that we might miss humour when nobody laughs. For further details, please see Kobayashi and Berge (2022).

We could not imagine any of the supervisors in the four cases expressing the same nervousness when suggesting something. This is typical for our data set that doctoral scholars' suggestions are accompanied with giggle or laughter, and the giggle gives the whole statement some ambiguity: Is it a joke or not? In this way, the doctoral scholars open up for the supervisors to interpret statements like this as jokes that are possible to ignore and move on. We also see a pattern that the doctoral scholars made nervous laughter when they tried to take up supervisors' time with details. This laughter could be interpreted as a hierarchical order in the room and humour becomes a 'safe way' to make suggestions (Berge, 2017), and making suggestions is a way to build independence.

In a related example doctoral scholar Sara laughed with insecurity as she touched upon the difficult issue of how to share data:

Supervisor:	So in your first paper, what do you intend to include in that, specifically?
DS Anish:	the body proportion, measurements and feed intake. And physical activity, I am analysing this.
Supervisor:	(writing down) physical activity, yes.
DS Anish:	and organ ways
Supervisor:	and what?
DS Anish:	Organ ways
Supervisor:	organ ways, yeah
DS Sara:	except the intestine **hehehehi**
Supervisor:	yeah, obviously.
DS Sara:	**hehehe** yeah

Case 3

The topic here is how to share data in a fair way. Anish explained what data he will use, and Sara added what data he should not use, with laughter to make it less harsh. When the supervisor agreed with her, she laughed again. Here laughter has the function of releasing tension and thereby making it easier to voice concerns and opinions.

The supervisors, on the other hand, used humour to soften harsh messages, as described by Li and Seale (2007). In the following excerpt the supervisor wants to have control:

Main supervisor:	that's very nice. So you're working on the challenge test now?
DS Anish:	yeah, and I also saw that some of the results (?)
Main supervisor:	yeah, but I am not supposed to see them or what? Because I am very curious **hehehe**
DS Anish:	**heh** The problem is that that [unclear] to make a graph.
DS Sara:	yeah, make a decent graph, right? **heha**
Main supervisor:	yeah. Because I am very curious.
DS Anish:	okay **heheheh**

Case 3

The supervisor here refers to the norm of calling—having a passion for research (Anderson et al., 2010), saying that she is 'very curious' as a reason for wanting to see the data, but our interpretation is that she wanted control. In addition, the doctoral scholars get an opportunity to learn about the norm of passion for research (Anderson et al., 2010). Here is an additional example from another session:

Supervisor:	Good. I'll do that… [*Unclear]* But I look forward to seeing a nice diagram of what you will do with the samples **HeHe**

Case 4

The supervisor sets high expectations, but softens this with humour. In our interpretation the laughter signals a room for failure and thereby creates a safe space for the doctoral scholar. Achievement goals are encouraging and support the doctoral scholar's development of independence as long as trust is the fundamental 'seedbed' for support (Devos et al., 2015).

Supervision is an opportunity to learn about norms and standards of science:

Co-supervisor:	I think that's a good idea
Main supervisor:	ja
Co-supervisor:	sometimes we are too ambitious
Main supervisor:	sometimes?
Co-supervisor:	**hehehe**

Case 1

In the preceding interaction they have lowered their first ambitions and admitted that they aimed too high before. The main supervisor then asks teasingly if they only sometimes aim too high. This humour touches on the norm of aiming for highest possible quality in research (Anderson et al., 2010), but at the same time managing feasibility. Although doctoral scholar Anne does not say anything here, she overhears the supervisors joking about being ambitious and at the same time realistic. This is a learning opportunity since she has to learn to judge her own level of ambitions against time limits in the process of becoming a more independent researcher.

In the following excerpt the doctoral scholar Cecilie took on the role of the supervisor when she was actually 'teaching' the postdoc:

Postdoc Kristian:	Then I can look them up on the micro ray and hopefully they are all then going for example up. If some are going up and some are going down/
DS Cecilie:	That's my next question then
Postdoc Kristian:	Yeah
DS Cecilie:	They should all kind of follow each other.
Postdoc Kristian:	Right. So this is what I have to do.
DS Cecilie:	Quality control **Hen**
Postdoc Kristian:	If it makes sense, yeah?
Supervisor:	hmmm (acknowledging)

Case 4

Cecilie pinpointed the important standard of quality control in good research. She is manifesting independence: she not only knows about these norms but can navigate them. We interpret the laughter and humour here as having the function of making her role less authoritarian since she is not the supervisor.

Our last excerpt shows how the two supervisors help the doctoral scholar Aisha in voicing her doubts:

Main supervisor:	but that didn't answer your question, I can see/
DS Aisha:	**hehe** no
Main supervisor:	**haha** I can see that from your face **hahaha**
Co-supervisor:	**haha**
	[all laughing] **hehehe hahahehe**
DS Aisha:	that's the idea
Main supervisor:	then you are sitting there: 'oh, what are they talking about?, I, I'
Co-supervisor:	**hahaha** but we agree very much **hahahaha**
Main supervisor:	yes, sure!
Co-supervisor:	OK, so what's the [your] question again?
Main supervisor:	**hahaha.** Try and phrase that again and we will see if we can
DS Aisha:	yes. How should I include this perfect idea **hehe** in my PhD plan **hehehe**
Main supervisor:	**hahahaha**
DS Aisha:	**hehe,** that is the question. I mean, yes I see, it's like scaling up, right?
	[...supervisors explaining...]
Main supervisor:	was that better?
DS Aisha:	yes **hehehehe**
Main supervisor:	**hme hme he**

Case 2

In this interaction the two supervisors realise that Aisha is confused by their answers but reluctant to say that aloud. They encourage her with humour to repeat her question, so as to ensure that she understands how to go about with the thesis. Aisha is answering with humour, talking about the supervisor's 'perfect idea' in her PhD plan. The main supervisor explains scaling up results from the lab to estimate implications for agriculture, which refers to the norm of doing relevant research (Kobayashi & Emmeche, 2023). The main supervisor uses humour here to make the situation a safe space, a space where you can laugh and ask 'stupid questions' as well as come up with stupid suggestions (Berge, 2017). This bears importance for her developing competent autonomy; she is not only supposed to do as the supervisors say, but also understand why. The supervisors create a fundamental ground for the possibility of voicing doubt in supervision, a necessity for good communication.

Discussion and Recommendations

Humour and laughter have important functions in supervision and doctoral education, and as we have seen in the examples above, doctoral scholars use humour to make suggestions and supervisors use humour to soften critical questions, or even to point out that the doctoral scholar does not understand (Case 2). We know from previous research that people of equal power will use more humour and different types of humour than those who have a very hierarchical relationship (Dunbar et al., 2012). Thus, humour can be a sign of equity. However, a joke is different depending on who is saying it in supervision—which is easy to forget (Li & Seale, 2007). Humour (and laughter) in itself does not make the relationship less hierarchical, but we believe that humour enables the doctoral scholar to have more courage, to voice doubt in a safe way, crossing the lines of power relations, and to take charge, as our examples in this chapter illustrate. Humour can be an agent in the hidden curriculum for supporting autonomy development when used in a constructive way. As we have seen, humour can help supervisors in their role as critical friends, when humour is used in an inclusive way that enables a safe space.

It is important to note that not all humour is positive: humour can function as a way to exclude others, for instance, when a group laughs at the same joke but those who do not understand the joke are excluded. Supervisors need to be aware that humour is a riskful form of communication and it can be awkward if one's sense of humour is not shared. There is also negative humour that causes harm (Billig, 2005), where we laugh down at others and thereby belittle them. The humour of racism and the humour of sexism are other examples of harmful humour. We hope that this chapter can contribute to more awareness, so that humour can be used more intentionally to build researcher independence.

We would like to round the chapter off with some pieces of advice. A basic understanding of how humour functions and what to be wary of is the first step in using humour and laughter intentionally and constructively. Then, the next step is to create a warm, friendly, informal atmosphere that invites humour and laughter. This may be easier said than done, humour is complex and subtle, and that is why humour is 'notoriously hard to translate' (Kuipers, 2009) and the last thing you learn when living in another culture. Generally doctoral scholars thrive better in informal settings, also across cultures (Elliot & Kobayashi, 2019). It may be a good start if you, as a supervisor, are able to laugh at yourself, using laughter as a sign of imperfection, to create a safe space in supervision and to craft a more relaxed way to discuss and look at things. This way you show that there is not one perfect way to become a professor. For example, Nick Hopwood emphasises the importance of imperfection and humour in his blog (Hopwood, 2017), where he writes about his 'rejection wall'. He makes a point of sharing his rejections and wishes that it would be commonplace in academia to say, 'You won't believe how awful the review I got this morning was! Come and laugh at it with me over coffee!'

Declarations of Conflicts of Interest We do not have any conflict of interest.

References

Anderson, M., Ronning, E., De Vries, R., & Martinson, B. (2010). Extending the Mertonian norms: Scientists' subscription to norms of research. *The Journal of Higher Education, 81*(3), 366–393.

Berge, M. (2017). The role of humor in learning physics: A study of undergraduate students. *Research in Science Education, 47*(2), 427–450. https://doi.org/10.1007/s11165-015-9508-4

Billig, M. (2005). *Laughter and ridicule: Towards a social critique of humour*. Sage.

Devos, C., Van der Linden, N., Boudrenghien, G., Azzi, A., Frenay, M., Galand, B., & Klein, O. (2015). Doctoral supervision in the light of the three types of support promoted in self-determination theory. *International Journal of*

Doctoral Studies, 10, 438–464. Retrieved from http://ijds.org/Volume10/IJDSv10p439-464Devos1858.pdf

Dunbar, N. E., Banas, J. A., Rodriguez, D., Liu, S. J., & Abra, G. (2012). Humor use in power-differentiated interactions.

Elliot, D. L., & Kobayashi, S. (2019). How can PhD supervisors play a role in bridging academic cultures? *Teaching in Higher Education, 24*(8), 911–929. https://doi.org/10.1080/13562517.2018.1517305

Gurr, G. M. (2001). Negotiating the 'rackety bridge'—A dynamic model for aligning supervisory style with research student development. *Higher Education Research and Development, 20*(1), 81–92. https://doi.org/10.1080/07924360120043882

Hopwood, N. (2017). *My wall of rejections and why it matters*. Blog on Academic work, publishing, writing, PhD stuff, qualitative methods. Retrieved from https://nickhop.wordpress.com/2017/06/21/my-wall-of-rejection-and-why-it-matters/

Jefferson, G., Sacks, H., & Schegloff, E. (1977). Preliminary notes on the sequential organization of laughter. *Pragmatics Microfiche, 1*, A2–D9.

Kobayashi, S., & Berge, M. (2022). Learning norms of science through laughter: A study of humour in life science supervision. *International Journal of Science Education*. https://doi.org/10.1080/09500693.2022.2090634

Kobayashi, S., & Emmeche, C. (2023). What makes a good PhD thesis? Norms of science as reflected in written assessments of PhD theses. Accepted for publication in *Assessment and Evaluation in Higher Education*.

Kuipers, G. (2009). Humor styles and symbolic boundaries. *Journal of Literary Theory, 3*(2), 219–239. https://doi.org/10.1515/JLT.2009.013

Li, S., & Seale, C. (2007). Managing criticism in Ph.D. supervision: A qualitative case study. *Studies in Higher Education, 32*(4), 511–526. https://doi.org/10.1080/03075070701476225

Merton, R. K. (1942). A note on science and democracy. *Journal of Legal and Political Sociology, 1*(1 & 2), 115–126. Heonline.

Merton, R. K. (1957). Priorities in scientific discovery: A chapter in the sociology of science. *American Sociological Review, 22*(6), 635–659.

Plester, B. (2015). *The complexity of workplace humour*. Springer.

Finding Confidence in Writing: Doctoral Writing Groups

Cally Guerin and Claire Aitchison

Introduction

One key outcome of doctoral programmes is an individual's transition from a 'student' guided by supervisors/advisers to an autonomous, independent 'researcher'. Through explicit and implicit learning, writing groups can play a valuable part in shaping how doctoral scholars experience this transformation (Wilmot & McKenna, 2018). Ideally, writing groups create safe, collegial spaces in which to notice, articulate, and rehearse new scholarly identities during the transition to becoming a professional researcher and knowledge maker. Writing groups can also activate the hidden curriculum of doctoral education in useful ways through offering academic, personal, social, and psychological support (Elliot

C. Guerin (✉)
Australian National University, Canberra, ACT, Australia
e-mail: cally.guerin@anu.edu.au

C. Aitchison
University of South Australia, Adelaide, SA, Australia
e-mail: Claire.Aitchison@unisa.edu.au

D. L. Elliot et al. (eds.), *Developing Researcher Independence Through the Hidden Curriculum*,
https://doi.org/10.1007/978-3-031-42875-3_13

et al., 2020). More than unlocking some of the underlying principles of good research writing, doctoral writing groups are also valuable for developing social and cultural understanding between participants and for enhancing their well-being (Aitchison, 2009; Aitchison & Guerin, 2014; Guerin et al., 2013).

In an apparently contradictory learning process, independence can be developed through working with peers in writing groups. Participants ostensibly learn from their peers—through direct discussion and from observation—what research writing should look like, what is regarded as 'good' writing and the processes for producing it (usually, this is their primary aim initially). However, often less obviously at the outset, interactions within these groups facilitate sharing of ideas about possibilities for expressing emerging researcher identities. Successful experiments in performing those identities that are witnessed and affirmed by group members create positive experiences that build participants' confident belief in themselves as researchers. Because this validating recognition comes from peers like themselves rather than supervisors, participants gradually realise they can trust their own understanding of what is required to succeed in the doctoral space, becoming increasingly self-reliant and self-sufficient as autonomous researchers. Thus, independence grows out of working together.

Models of Writing Groups

In previous work (Aitchison & Guerin, 2021), we have identified two main types of writing groups: critiquing groups, in which members read text and offer feedback; and productivity groups, in which members get together to work on their own texts. These groups can be set up in whatever way best suits the specific members. Haas (2014) provides an extensive typology to guide these decisions, including whether to conduct meetings online or in person. Many doctoral scholars find it effective to engage in a combination of critiquing and producing during meetings—an hour focused on discussion of text followed by an hour of writing in company can work very well. Both critiquing and productivity groups can develop the independence of doctoral scholars through access to elements of the hidden curriculum, but do so in slightly different ways: critiquing groups are particularly adept at revealing the hidden curriculum related to conventions of collegial peer review; productivity groups are helpful for uncovering the hidden processes of how others get on with their writing.

Critiquing Groups

Writing groups that focus on critiquing members' work have been part of writers' study and support strategies for many years (Paré, 2014). In these groups, authors read each other's work and offer collegial, constructive feedback on the text. This feedback might take the form of written or verbal comments; it might be delivered during a real-time, synchronous meeting or it might be shared electronically and asynchronously. Whatever the format, these groups benefit from establishing clear expectations of what qualifies as 'constructive' critique: agreed guidelines can be invaluable in articulating shared expectations around content and behaviours. It is recommended that those guidelines are based on consideration of the following questions.

- *How much reading are participants willing/able to undertake?*

It's surprising how much discussion 1000–2000 words can generate. The specific examples put forward by participants often lead to more general conversation that clarifies and explores a range of related writing issues.

- *Will comments be framed around the text, rather than be expressed as 'You do/don't do this'?*

This distancing between the author and the manuscript helps writers focus on the communication rather than their own emotions about the text. Instead of feeling personally criticised by reviewers, the author can consider how the text can be improved to carry its message effectively.

- *When identifying an issue with the writing, do you expect the reviewer to offer a solution?*

This can be the most valuable part of the learning for those offering critique (Caffarella & Barnett, 2000; Lundstrom & Baker, 2009). It is one thing to notice some aspect of the writing isn't working, and quite another to work out how to fix it. This growing awareness of what makes for effective written communication feeds back into the reviewer's own writing.

- *Are all members expected to offer critique, even if they can't attend in person?*

Doctoral scholars are often very busy people, so there will inevitably be times when some group members are unable to attend meetings. Nevertheless, the success of writing groups is partly dependent upon everyone contributing and sharing knowledge and information about writing. Clearly, these groups will not achieve their goals if one scholar offers their work for critique, gathers all the feedback from others, but is then too busy to read and respond to the writing of other group members. A commitment to send comments in advance to the group if unable to attend in person can be helpful; those comments can then also inform the discussion.

- *Will writers have a 'right of reply' in defending their work?*

When writers have put so much time and energy into producing the document, a common impulse is to explain all their decisions and why the work is already very good. However, it can be very useful to encourage the writer to listen to the critique even if they feel it is wrong-minded—after all, this provides insight into the reader's experience of the text. Afterwards, authors can make up their minds about which advice to act upon.

Writers themselves can identify what aspects of their writing they would like the group to comment on. For example, they may be uncertain about the logic of the structure, persuasiveness of the argument, authoritative voice, suitably 'academic' style, word choice, clarity of expression, appropriate level of detail, grammatical correctness. Perhaps the writer has struggled with a particular aspect of the text, and would like the group to check whether they have solved the problem they had identified; or perhaps they are stuck and seek advice on how to deal with an aspect of the writing that is not effectively communicating their ideas.

The overt purpose of critiquing groups is to learn about how to improve one's own academic and research writing. However, in a clear example of how the hidden curriculum of doctoral education operates, participation in these groups also very directly teaches the skills of peer review. In offering critique of academic texts, members learn what elements of the writing they should pay attention to, what is expected in an academic text, and how to communicate both praise and criticism in helpful ways. Alongside this constructive feedback, members also learn the expectations of collegial, scholarly behaviour: how to evaluate writing, how to present critique, and how to participate respectfully in robust academic debate. The groups provide a training ground to practice these social skills and dispositions

(Cahusac de Caux et al., 2017; Sletto et al., 2020). In the process of developing peer review skills, participants begin to build confidence in their own ability to perform research well, and earn the trust of peers as they offer opinions on the research writing under scrutiny. Hidden just below the surface of writing group activity is the process of transformation into a researcher identity.

Productivity Groups

In production-focused writing groups, doctoral scholars arrange a time and place (online or in person) to meet and commit to getting on with doing some writing. This style of writing group became popularised through the 'Shut up and write!' model (Mewburn et al., 2014). These groups include some social time and some focused writing time, often based around a pre-determined schedule (e.g., a 2-hour session might meet at 10 am and chat, write for 50 minutes, have a 10-minute social break, then write for another 50-minute sprint).

Ideally, participants nominate their planned writing goal at the beginning of the session and then report back at the end on what they managed to achieve. This has a number of benefits:

- Writers start to develop a sense of how long writing actually takes. Many doctoral scholars are overly optimistic about how quickly they can produce text. By seeing the difference (or alignment) between their planned task and what they achieved, more realistic time frames can be implemented.
- This improved workload and time management has the added benefit of shifting attention from the frustration and impatience of not achieving planned outcomes to identifying progress on the writing.
- Observing what others nominate as reasonable writing tasks to achieve during the session also provides a benchmark for how much writing their peers are able to produce during that time. The individual nature of most doctoral projects means that it can be very difficult to measure whether a scholar is keeping up with their peers.
- As well as *how much* writing, publicly nominating tasks can provide insights into *what* others focus on. This can usefully highlight aspects of the writing that might otherwise be overlooked, or reveal approaches to writing that are focused on unimportant elements of the document.

Productivity groups provide access to particular elements of the hidden curriculum of doctoral education. Here participants learn about their own capacity to produce useful text on demand, avoiding procrastination and taking a disciplined approach to the work. Simultaneously, participants observe the routines and writing practices of peers. Doctoral scholars learn to take control of their own time and planning, noticing what they can achieve in a given timeframe and identifying their own success in making progress.

Independence and the Hidden Curriculum

Both kinds of writing groups build confidence and facilitate independence by activating the hidden curriculum of doctoral education; both provide learning opportunities related to academic, personal, social, and psychological aspects of doctoral education.

Critiquing groups are particularly useful in providing access to the academic elements of the hidden curriculum. In these spaces, doctoral scholars are invited to demonstrate their critical thinking, share innovative approaches to problem solving, and exhibit knowledge of their field. Interdisciplinary groups can stimulate valuable discussions about clarity in writing, and also—importantly—identify characteristics of research writing that are distinct to specific disciplinary contexts (Colombo & Rodas, 2020; Guerin et al., 2013). Participants learn to manage multiple voices as they make decisions about which feedback advice to adopt and which to reject (Aitchison, 2014). Equally, they realise they themselves have useful insights to pass on and start to regard themselves as valued, knowledgeable colleagues and experts.

What is often overlooked in formal programmes is how academic concerns interact with the research culture and administrative systems. It is here that insider knowledge shared between peers can be invaluable. For example, peers can offer up-to-date advice on who to talk to about paperwork; how to behave appropriately in seminars; extras that are available from the library; which researcher development workshops are good use of time. This sharing of information facilitates the independence of doctoral scholars as they become less reliant on supervisors to validate their work; instead, they can take agency in navigating the academic system, seeking help, advice, and insider knowledge from peers.

Productivity groups offer a path towards independence through a focus on the practices of doctoral writing. The academic learning in these groups tends to occur through participants observing how others approach the task of settling down and getting on with the work (Kumar & Aitchison, 2018). Monitoring goals and noting achievements in each session demonstrate the progress being made as scholars move towards 'doctorateness' (Trafford & Leshem, 2009).

A holistic approach to researcher development appreciates the influence of scholars' personal lives and social interactions on their doctoral studies. The work space of both types of writing group readily allows for conversation and friendship; they can also be a powerful influence on developing researcher identities (Danvers et al., 2019; Grant, 2006). Here, doctoral scholars can share valuable learning related to undertaking a PhD; the trust that is built between group members can also open a safe space to air personal concerns and share strategies for handling those challenges (e.g., how to manage childcare responsibilities, where to shop for more affordable groceries). This trust develops from the vulnerability that is inevitably exposed when sharing work in progress and inviting feedback and critique, or in stating ambitious writing goals that may or may not be achieved. Instead of seeing only the final, polished outcomes on display in a formal seminar presentation, the focus in writing groups is on the process of doing the work. Peers here can teach each other about what choices might be available to them and how to take control of those choices.

A major advantage of doctoral writing groups is the activation of collegial interactions between local and international doctoral scholars (Guerin et al., 2013). Peers meet as equals in this space, offering different perspectives and insights on writing processes. Everyone has the opportunity to use their knowledge and display their capacity, regardless of their first language and how experienced they are in writing about research in English. Writing groups thus create a space where doctoral scholars recognise each other as a valuable resource through experiences of learning from each other. They are no longer solely dependent on supervisors for advice and guidance. (See a similar experience described in Makara et al.'s chapter where doctoral scholars actively took part in a Journal Club.)

Finally, the confidence that is built through participation in doctoral writing groups is a clear psychological benefit (Guerin, 2014; Johnson, 2019). Interacting with productive peers and connecting with a supportive research community have been shown repeatedly to enhance wellbeing for doctoral scholars (Hradsky et al., 2022). The message becomes 'I can do this'. When comments on drafts are taken seriously and demonstratively improve the work, when writers observe good progress in producing documents, the identity as a legitimate scholar is positively reinforced. Gradually, the transformation into an autonomous researcher occurs.

Paradoxically, this connection into the research community through writing groups is a valuable pathway to independence: doctoral scholars build an identity as a discrete self who collaborates and cooperates with peers. Opportunities to learn how to conduct oneself effectively in the doctoral research space result in self-reliant and self-sufficient doctoral scholars, who are no longer solely dependent on their supervisors to access the hidden curriculum of doctoral education. Here, 'independence' means moving autonomously amongst peers.

Declarations of Conflicts of Interest There are no conflicts of interest to declare.

Acknowledgements We acknowledge the hundreds (possibly thousands) of doctoral writers whose participation in writing groups has contributed to our understanding of this topic.

References

Aitchison, C. (2009). Writing groups for doctoral education. *Studies in Higher Education, 34*(8), 905–916. https://doi.org/10.1080/03075070902785580

Aitchison, C. (2014). Learning from multiple voices: Feedback and authority in doctoral writing groups. In C. Aitchison & C. Guerin (Eds.), *Writing groups for doctoral education and beyond: Innovations in practice and theory*. Routledge.

Aitchison, C., & Guerin, C. (Eds.). (2014). *Writing groups for doctoral education and beyond: Innovations in practice and theory*. Routledge.

Aitchison, C., & Guerin, C. (2021). *Designing a research writing group: What leads to success?* Presentation for Pedagogical devices for the development of research writing SIG (Bogota, 11/12 Feb).

Caffarella, R. S., & Barnett, B. G. (2000). Teaching doctoral students to become scholarly writers: The importance of giving and receiving critiques. *Studies in Higher Education, 25*(1), 39–52. https://doi.org/10.1080/030750700116000

Cahusac de Caux, B. K. C. D., Lam, C. K. C., Lau, R., Hoang, C. H., & Pretorius, L. (2017). Reflection for learning in doctoral training: Writing groups, academic writing proficiency and reflective practice. *Reflective Practice, 18*(4), 463–473. https://doi.org/10.1080/14623943.2017.1307725

Colombo, L., & Rodas, E. L. (2020). Interdisciplinarity as an opportunity in Argentinian and Ecuadorian writing groups. *Higher Education Research & Development, 40*(2), 207–219. https://doi.org/10.1080/07294360.2020.1756750

Danvers, E., Hinton-Smith, T., & Webb, R. (2019). Power, pedagogy and the personal: Feminist ethics in facilitating a doctoral writing group. *Teaching in Higher Education, 24*(1), 32–46. https://doi.org/10.1080/13562517.2018.1456423

Elliot, D. L., Bengtsen, S. S., Guccione, K., & Kobayashi, S. (2020). *The hidden curriculum in doctoral education*. Springer Nature.

Grant, B. M. (2006). Writing in the company of other women: Exceeding the boundaries. *Studies in Higher Education, 31*(4), 483–495. https://doi.org/10.1080/03075070600800624

Guerin, C. (2014). The gift of writing groups: Critique, community and confidence. In C. Aitchison & C. Guerin (Eds.), *Writing groups for doctoral education and beyond: Innovations in practice and theory*. Routledge.

Guerin, C., Xafis, V., Doda, D. V., Gillam, M., Larg, A., Luckner, H., Jahan, N., Widayati, A., & Xu, C. (2013). Diversity in collaborative research communities: A multicultural, multidisciplinary thesis writing group in Public Health. *Studies in Continuing Education, 35*(1), 65–81. https://doi.org/10.1080/0158037X.2012.684375

Haas, S. (2014). Pick-n-Mix: A typology of writers' groups. In C. Aitchison & C. Guerin (Eds.), *Writing groups for doctoral education and beyond: Innovations in practice and theory*. Routledge.

Hradsky, D., Soyoof, A., Zeng, S., Foomani, E. M., Cong-Lem, N., Maestre, J.-L., & Pretorius, L. (2022). Pastoral care in doctoral education: A collaborative autoethnography of belonging and academic identity. *International Journal of Doctoral Studies, 17*, 1–23. https://doi.org/10.28945/4900

Johnson, E. M. (2019). Towards an enhanced view of doctoral writing environments: Learning alliances to reconceptualise practice. *Policy Futures in Education, 17*(2), 140–152. https://doi.org/10.1177/1478210318774441

Kumar, V., & Aitchison, C. (2018). Peer facilitated writing groups: A programmatic approach to doctoral student writing. *Teaching in Higher Education, 23*(3), 360–373. https://doi.org/10.1080/13562517.2017.1391200

Lundstrom, K., & Baker, W. (2009). To give is better than to receive: The benefits of peer review to the reviewer's own writing. *Journal of Second Language Writing, 18*(1), 30–43. https://doi.org/10.1016/j.jslw.2008.06.002

Mewburn, I., Osborne, L., & Caldwell, G. (2014). Shut up & Write! Some surprising uses of cafés and crowds in doctoral writing. In C. Aitchison & C. Guerin (Eds.), *Writing groups for doctoral education and beyond: Innovations in practice and theory*. Routledge.

Paré, A. (2014). Writing together for many reasons. In C. Aitchison & C. Guerin (Eds.), *Writing groups for doctoral education and beyond: Innovations in practice and theory*. Routledge.

Sletto, B., Stiphany, K., Futrell Winslow, J., Roberts, A., Torrado, M., Reyes, A., Reyes, A., Yunda, J., Wirsching, C., Cho, K., & Tajchman, K. (2020). Demystifying academic writing in the doctoral program: Writing workshops, peer reviews, and scholarly identities. *Planning Practice & Research, 35*(3), 349–362. https://doi.org/10.1080/02697459.2020.1748331

Trafford, V., & Leshem, S. (2009). Doctorateness as a threshold concept. *Innovations in Education and Teaching International, 46*(3), 305–316. https://doi.org/10.1080/14703290903069027

Wilmot, K., & McKenna, S. (2018). Writing groups as transformative spaces. *Higher Education Research & Development, 37*(4), 868–882. https://doi.org/10.1080/07294360.2018.1450361

Doctoral Intelligence Mechanisms to Illuminate Development Strategies in the Hidden Curriculum

Ruth M. Albertyn

Introduction

> *...all this gymnastics makes you resilient and it advances your level of abstraction ... the journey of the PhD makes you develop this skill.*

This comment by a doctoral graduate reveals unique learning during the doctoral process. This 'advanced level of abstraction' is referred to in the conceptualisation of 'doctorateness' and reflects higher levels of thinking and quality research (Trafford & Leshem, 2008; Yazdani & Shokooh, 2018). There are increasing calls for innovative doctoral education to develop independent researchers, who produce this quality research and contribute to the society beyond the qualification (Nerad et al., 2022). Generic training programmes have been criticised as they focus on

R. M. Albertyn (✉)
Centre for Higher and Adult Education/Stellenbosch Business School, Stellenbosch University, Stellenbosch, South Africa
e-mail: rma@sun.ac.za

D. L. Elliot et al. (eds.), *Developing Researcher Independence Through the Hidden Curriculum*,
https://doi.org/10.1007/978-3-031-42875-3_14

efficiency and the research product, sometimes at the cost of quality, and not on holistic development during the PhD (Torka, 2018). Bengtsen and McAlpine (2022) noted that supervisors were more likely to give students advice on 'how to'—on the instrumental level of development—rather than on risk-taking and finding their own voice as researchers. Elliot et al. (2020) concur and note that educational developers often guide towards the PhD product but not towards independence. We may thus ask what we can learn about the learning mechanisms that may be hidden during the doctoral process and how we foster this development based on doctoral scholars' experiences.

The aim of this chapter is to propose a doctoral intelligence (DI) framework for research independence through the hidden curriculum based on evidence of scholars' development during the doctorate. This DI framework includes hidden curriculum dimensions and principles for fostering researcher independence that could act as a dynamic anticipatory, diagnostic, and development tool to guide research development.

Doctoral Intelligence Conceptualisation

The term 'intelligence' indicates the inherent knowledge and abilities for accomplishing a task, or those mind-sets developed and employed for problem solving in a specific context (Nisbett et al., 2012; Sternberg, 2000). The development perspective of intelligence assumes that both internal and external factors contribute to intelligence (Davidson & Downing, 2000). In the doctoral context, internal factors relate to the individual's innate cognitive skills and attributes that qualify them to embark on doctoral studies. External factors refer to environmental aspects that can facilitate development of intelligence such as a range of doctoral education interventions and support. Intelligence is thus seen to be dynamic and can change as the environment changes (Nisbett et al., 2012). Sternberg further asserts that there are not defining attributes, but only characterising attributes that tend to be typical of intelligent persons. Therefore, the need for flexibility when defining intelligence for problem solving valued in a particular context (Earley et al., 2006), in the case of this chapter, defining doctoral intelligence for the doctoral context.

A cornerstone of innovative doctoral education could be to enhance mind-sets necessary to promote doctorateness. DI has been conceptualised to indicate the 'knowing', 'doing', 'thinking', and 'willing' mind-sets

Textbox 1: *DI domains*

Knowing:	Foundational expertise in discipline and research
Doing:	Application in practice of research for the PhD product
Thinking:	Higher level mental processing for quality doctoral work
Willing:	Open-minded for continuing development

(Textbox 1) for completing the doctorate based on an analogous link to cultural intelligence (see Albertyn, 2021; Earley et al., 2006).

The focus of DI is thus on conceptualisation of characterising mind-sets and not on the multiple competencies reported by Durette et al. (2016), for example. Mowbray and Halse (2010) believe that such lists of competencies may be daunting. Therefore, a focus on broad mind-sets could lighten the cognitive burden and enlighten doctoral scholars about expectations in a way that is empowering. This chapter builds on earlier DI conceptualisation and empirical work reporting evidence of the four domains but focuses specifically on hidden DI domains and principles identified that could guide doctoral pedagogy.

Doctoral Intelligence Manifestations Reflect Mechanisms for Dynamic Development

Experiences of doctoral education were explored during interviews with questions related to the DI domains of 'knowing', 'doing', 'thinking', and 'willing'. Purposive sampling was applied to select twenty-two doctoral scholars from three PhD programmes (Development finance, Futures studies, and Business management and administration) at four stages of studies (proposal, implementation, concluding, and graduated). Selective evidence from findings reflects hidden curriculum DI manifestations and mechanisms and culminate in a DI framework to guide development efforts.

Doctoral Intelligence Manifestations

Manifestations of the four DI domains lead to a clearer understanding of mechanisms that lead to independence during the learning process.

Knowing

It would be expected that scholars have expert knowledge of their disciplinary field and of research method in pursuit of this PhD qualification—referred to as 'grasp' by Holbrook et al. (2015). Analysis revealed other 'knowing' aspects in the hidden curriculum related to embracing depth of knowledge in the PhD process that leads to novelty: 'That drive to really understand things deeply… the PhD gave me the tools and confidence to do that in a different way… connecting the dots and understanding the relationships'.

There is evidence of knowing as a process but also as a basis for lifelong learning. Beyond knowledge required for the PhD product, personal value and acknowledging the contribution to society were noted:

> It is adding tremendous value to my life… I am learning a lot about the discipline… I am learning about myself
> … being comfortable with engaging with those different fields and… find a way for all of those fields to live within my study and then a lot deeper and it feels a lot more responsible in terms of that whatever you are saying here, it better be meaningful.

Another participant noted self-knowledge: 'I have to teach myself how to learn this myself'. These comments reflect the alternative types of knowing that are hidden during the process while learning for independence during the PhD.

Doing

The application phase is essential for completing the PhD research product; but the additional benefits relate to the confidence obtained through learning by doing. Their independence needs to be demonstrated through informed research decision making for application that provides them with validation (Ashforth & Schinoff, 2016). Thus, elements of ownership and agency develop through doing the work required to attain the PhD:

> you had to create your own authentic structure and if it is not genuine, it will show, people will pick up … company[ies] want people like that, people who can work by themselves. … You can think, process, and create more. … They like people who can solve problems.

Doing PhD research creates an impetus for learning with reciprocal benefits on the product and identity development and independence for the doctorate and beyond the qualification.

Thinking

The thinking domain reflects higher-order mental processes and level of cognitive functioning (Davidson & Downing, 2000). This domain is often a more challenging part of the doctorate (Trafford & Leshem, 2008). Aligned with literature that refers to critical and creative thinking as being crucial in doctoral education (Brodin & Frick, 2011; Hodgson, 2020), scholars in this study also referred to independence in thinking and being able to crisply communicate thinking. As indicated by one participant: 'I actually loved that ... it is for me to decide ... I need to take full responsibility and control of that thought process'. As participants were mainly mature part-time scholars, they reported using divergent and convergent thinking problem solving tools they use in the workplace. Value was found in diverse activities, such as interactions with others in the workplace or in colloquia where their thinking was challenged or affirmed. This development built confidence and depth and contributed to the quality of the PhD, leading to independence and formation of their research identities.

Willing

The willing mind-set seems to be foundational throughout the doctoral journey. It also develops through the process of learning. This mind-set is reflected by one graduate: 'you are very angry and depressed, but that ability to come back and try to see what is the point here ... it is humility. You have to be humble.' Aligned with the attributes of intellectual virtues proposed by Ortwein (2015), participants identified the following mind-sets: responsibility, purpose, curiosity, being open-minded, love of learning, humility, excellence, mental maturity, and wisdom. They indicated that, in addition to motivation, there needs to be greater purpose or value to drive the continual process of development and prevent scholars from giving up: 'if there is some sort of an external contribution. If you have it, you do not even see it as a really long journey. ... The iterative nature and various nuances of the doctoral process harness this development and sustain the learning process for quality products valuable in society'.

Doctoral Intelligence Mechanisms

Three DI mechanisms reveal principles that could guide the hidden curriculum, namely, embracing the learning during the doctoral process, encouraging strategies for ownership and independence, and harnessing integrated DI domains as part of a dynamic iterative process for development towards independence.

Learning During the Doctoral Process

> you are building up, building up. … Steadily climbing steps. It is like Great Wall of China. … You cannot go quicker. … So, I find that I am more deliberate … you are thinking in a more consistent pattern.

This comment about learning due to the doctoral process was mentioned by others also who referred to the environment 'pushing' scholars due to expectations by the context, the nature, and outcomes of the qualification and by society. This process is not always pleasant and forces reflection and deeper engagement: 'a lot of circling around in my mind'. These scholars refer to being 'rigorous with themselves' due to evidence needed to give 'surgical sharpness' to arguments. Higher levels of abstraction were developed due to the continual iterative exploration in research itself facilitating this learning rather than abstraction being 'taught'. Graduates indicated how the PhD prepared them for problem solving in the world in general and the continued curiosity and engagement after qualification indicating skill retention.

One comment, 'At the beginning I was very confused. I am now less confused. Not unconfused yet', reflects this scholar making peace with this process of learning. Thus, scholars need to gain insight into the nature and value of learning as a trajectory of development during the doctoral process without over-focusing on progression towards the PhD product. This principle reflects the integration of the DI domains and reveals the hidden curriculum dimensions beyond disciplinary and research knowledge and application.

Strategies for Ownership and Independence

The reported strategies scholars employed reveal their independence and agency (see Table 1).

Table 1 Hidden strategies for development towards research independence

Structure	Putting systems in place for accountability; to identify tools and resources; to simplify and understand the essence first by breaking things 'down in chunks'
Alignment	Continual checking of alignment and the golden thread by keeping the research question central, to avoid getting side-tracked: 'if you engage with it continuously, you start connecting things'
Brainstorming	Talking to others to 'test the waters' and listening to others to get ideas. Mind-mapping/relevance tree/drawing of conceptual maps to connect the dots and seeing things from a different perspective. Keeping a researcher notebook for ideas as they arise
Distancing	Take a break, sleeping it off, slowing down the mind by being active [walking, swimming, or cycling]. Benchmarking, wide reading: 'and then a lot of stepping away and then just say, okay, let us just render this for a while'. As another participant noted: 'Rather down tools and not sit and mope and muddle, because while I am not here, the brain is working'
Distillation	The summarising and draft-upon-draft process helps with filtering to focus. 'Be comfortable with deleting'. Another tool mentioned was the three-minute speech. One person referred to simplicity through drawing and 'drilling down' into the key concepts
Personal investment	Self-talk, self-management, and ownership: 'You need to really interrogate issues on your own.' Believing in the larger purpose and responsibility to society was indicated, such as: 'how can I make a difference?' and 'to always have that appetite of resolving complex problems in society'

These six strategies reveal the hidden learning that takes place during the doctorate and reflects the integrations with the other DI domains. These strategies could act as a guide for research development towards independence. Scholars bring their own unique skills to the educational process and encouraging them to use these contributes to confidence, identify development, empowerment, and independence.

Integrating DI Domains Enhances Development

The mind-sets for success seem to be integrated during continuing learning in both the formal and hidden curriculum. (See also Elliot's chapter exploring the role of mind-set in enabling doctoral scholars to appreciate the value embedded in their experience.) The 'willing' DI domain seems to be a foundational domain infused throughout the PhD and a necessary mind-set for an incremental development (Wang et al., 2019). The attribute of intellectual humility within the willing domain seems to be crucial. Intellectual humility involves a realistic appraisal of strengths and weaknesses (Haggard et al., 2018) and has been linked to cognitive flexibility (Zmigrod et al., 2019), open-mindedness (Porter & Schumann, 2018), and wisdom (Wang et al., 2019). Principles related to learning through the process, ownership, independence, and integrating DI domains could foster dynamic development efforts towards researcher independence.

Hidden Curriculum Guiding Framework for Research Independence

Based on examples of hidden curriculum manifestations and mechanisms, the proposed DI framework could foster research independence development (see Fig. 1).

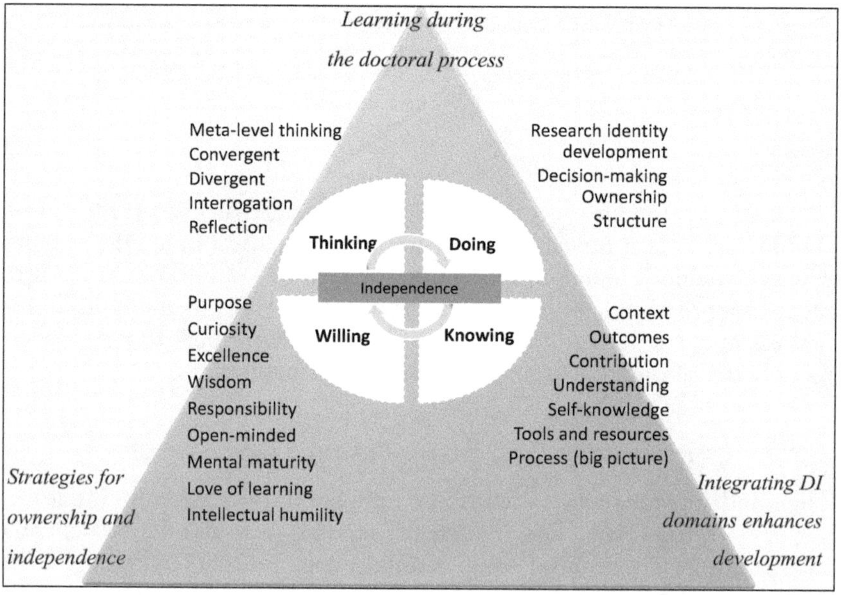

Fig. 1 DI framework for research independence through the hidden curriculum

This framework provides an overview of the manifestations of the hidden curriculum within the four DI domains and the three identified principles or mechanisms for development. Although the 'knowing' and 'doing' domains are traditionally focused on informal education programmes, this study revealed hidden indicators within these domains that influence development during the doctorate. These doctoral mindsets do not develop linearly and cannot be addressed with one-time training or tracked using competency checklists. Researcher independence is a dynamic process that develops iteratively over time.

Conclusion

There seems to be evidence that the DI mind-sets are not neatly packaged into a sequential set of steps that can be covered in formal educational programmes. The DI mind-sets evident in the hidden curriculum provide insights into pertinent mechanisms and principles that could influence thinking and action in doctoral pedagogy for continual research development infused in supervision, education, and support.

The DI Framework for Research Independence through the hidden curriculum provides a map of the doctoral terrain and could act as an anticipatory system reflecting dynamic mind-sets (Slaughter, 2008). In this holistic view of DI, the mind-sets present in one domain may influence the mind-sets in the other. This framework may also act as a diagnostic tool for discerning where one aspect is dominating other domains (Wilber, 2005). Due to the reciprocal relationship between domains, thoughtful strategies to ensure development of each domain could enhance development in the other domains. This meta-perspective of mind-sets provides guidance for pedagogical approaches for facilitating researcher independence during the doctorate and thereafter.

References

Albertyn, R. (2021). Making a case for Doctoral Intelligence: Conceptualisation and insights for researcher development. *Innovations in Education and Teaching International*. https://doi.org/10.1080/14703297.2021.1899033

Ashforth, B. E., & Schinoff, B. S. (2016). Identity under construction: How individuals come to define themselves in organizations. *Annual Review of Organizational Psychology and Organizational Behavior, 3*(March), 111–137.

Bengtsen, S. S., & McAlpine, L. (2022). A novel perspective on doctoral supervision interaction of time, academic work, institutional policies, and life course. *Learning and Teaching, 15*(1), 21–45. https://doi.org/10.3167/latiss.2022.150103

Brodin, E. M., & Frick, L. (2011). Conceptualizing and encouraging critical creativity in doctoral education. *International Journal for Researcher Development, 2*(2), 133–151. https://doi.org/10.1108/17597511111212727

Davidson, J. E., & Downing, C. L. (2000). Contemporary models for intelligence. In R. J. Sternberg (Ed.), *Handbook of intelligence* (pp. 34–52). Cambridge University Press.

Durette, B., Fournier, M., & Lafon, M. (2016). The core competencies of PhDs. *Studies in Higher Education, 41*(8), 1355–1370. https://doi.org/10.1080/03075079.2014.968540

Earley, P. C., Ang, S., & Tang, J. (2006). *Developing cultural intelligence at work.* Stanford University Press.

Elliot, D. L., Bengtsen, S. S. E., Guccione, K., & Kobayashi, S. (2020). *The hidden curriculum in doctoral education.* Springer. https://doi.org/10.1007/978-3-030-41497-9

Haggard, M., Rowatt, W. C., Leman, J. C., Meagher, B. R., Moore, C., Fergus, T., Whitcomb, D., Battaly, H., Baehr, J., & Howard-Snyder, D. (2018). Finding middle ground between intellectual arrogance and intellectual servility: Development and assessment of the limitations-owning intellectual humility scale. *Personality and Individual Differences, 124*, 184–193.

Hodgson, D. (2020). Helping doctoral students understand PhD thesis examination expectations: A framework and a tool for supervision. *Active Learning in Higher Education, 21*(1), 51–63. https://doi.org/10.1177/146194679878471471774422020

Holbrook, A., Bourke, S., & Fairbairn, H. (2015). Examiner reference to theory in PhD theses. *Innovations in Education and Teaching International, 52*(1), 75–85. https://doi.org/10.1080/14703297.2014.981842

Mowbray, S., & Halse, C. (2010). The purpose of the PhD: Theorising the skills acquired by students. *Higher Education Research and Development, 29*(6), 653–644.

Nerad, M., Bogle, D., Kohl, U., O'Carroll, C., Peters, C., & Scholz, B. (2022). Guiding principles. In M. Nerad, D. Bogle, U. Kohl, C. O'Carroll, C. Peters, & B. Scholz (Eds.), *Towards a global core value system in doctoral education* (pp. 43–50). UCL Press.

Nisbett, R. E., Aronson, J., Blair, C., Dickens, W., Flynn, J., Halpern, D. F., & Turkheimer, E. (2012). Intelligence: New findings and theoretical developments. *American Psychologist, 76*(2), 130–159.

Ortwein, M. J. (2015). The regulation of understanding through intellectual virtue: Some implications for doctoral education. *Journal of Thought, 49*(1–2), 71–85.

Porter, T., & Schumann, K. (2018). Intellectual humility and openness to the opposing view. *Self and Identity, 17*(2), 139–162.

Slaughter, R. (2008). What difference does integral make? *Futures, 40*(2), 120–137.

Sternberg, R. J. (2000). The concept of intelligence. In R. J. Sternberg (Ed.), *Handbook of intelligence* (pp. 3–16). Cambridge University Press.

Torka, M. (2018). Projectification of doctoral training? How research fields respond to a new funding regime. *Minerva, 56*(2), 59–83. https://doi.org/10.1007/311024-018-9342-8

Trafford, V., & Leshem, S. (2008). *Stepping stones to achieving your doctorate: Focusing on your viva from the start.* Open University Press.

Wang, J., Yang, X., & Fudan, J. (2019). Intellectual humility and owning one's limitations. *Journal of the Humanities and Social Sciences, 12*(3), 353–369.

Wilber, K. (2005). Introduction to integral theory and practice IOS basic and the AQAL map. *AQAL Journal of Integral Theory and Practice, 1*(1), 1–38.

Yazdani, S., & Shokooh, F. (2018). Defining doctorateness: A concept analysis. *International Journal of Doctoral Studies, 13*(January), 31–48. https://doi.org/10.28945/3939

Zmigrod, L., Zmigrod, S., Rentfrow, P. J., & Robbins, T. W. (2019). The psychological roots of intellectual humility: The role of intelligence and cognitive flexibility. *Personality and Individual Differences, 141*, 200–208.

Enabling Part-Time Doctoral Scholars to Develop Effective Support Villages

Jon Rainford

Introduction

For part-time doctoral scholars, the hidden curriculum can feel not just hidden but locked away from view with no keeper of the keys in sight. Proximity to other doctoral scholars and 'hidden curriculum agents' can provide the keys to the hidden curriculum through 'informal interactions' (Elliot et al., 2020, p. 6) and yet the nature of the part-time doctorate often limits these interactions, thus hiding the keys. Corridor conversations are a prime example of valuable informal interactions (Elliot et al., 2020) but what about when the corridors are removed? To facilitate these, this chapter argues that part-time doctoral scholars need to develop a support village or multiple support villages where these interactions can take place and enable them to find the keys they need. Whilst we can exist in isolation, it is often the people around us who provide support and companionship and help us make sense of the world around us.

J. Rainford (✉)
The Open University, Milton Keynes, UK
e-mail: jon.rainford@open.ac.uk

D. L. Elliot et al. (eds.), *Developing Researcher Independence Through the Hidden Curriculum*,
https://doi.org/10.1007/978-3-031-42875-3_15

It is no surprise that during competitions, athletes also live in villages. Even in the most individual sports such as long-distance running, a whole range of other people contribute to success, including coaches, nutritionists, physiotherapists, and many others. Like any good athlete, doctoral scholars also need a team around them to ensure they are given the best chances of performing well (Mantai, 2020; Rainford, 2021; Rainford & Guccione, 2023). For the doctoral scholar, however, this notion of a village might be more metaphorical than an athlete's support village and equally may be less likely to be constrained by place. Through foregrounding what makes a valuable support village, researcher developers can help improve access to the hidden curriculum through supporting part-time doctoral scholars to build their own support villages. Whilst the notion of a support village is not unique to part-time doctoral scholars, they become even more important when the proximity to other doctoral scholars and institutional support mechanisms is removed by this mode of study.

What Is Different for Part-Time Doctoral Scholars?

There are numerous forms of part-time doctorate which vary in the opportunities provided for proximity to other doctoral scholars (Rainford & Guccione, 2023). In the case of taught doctorates, there may be a cohort element for example, but this tends to be focused upon short, condensed interactions, often with peers at the same stage of their journey. This is unlike the frequency with which full-time doctoral researchers are likely to encounter peers and other hidden curriculum agents. They are also likely to encounter fewer different agents when the limited on-campus contact is likely to be focused on academic teaching staff. Furthermore, chance encounters are less likely, so more planning and scheduling are needed to create these connections, thus slowing down access to this support and impacting the informal support available.

Demystifying the Belief of the Lone Scholar

One of the often-forgotten steps in the journey to researcher independence is the value of interdependence to help scholars transition to a position of increased researcher agency. As Elliot et al. (2020, p. 134) have argued, 'agency is essential to thriving as a doctoral researcher', yet there is an inherent danger that agency is interpreted as going alone on the journey as opposed to taking control of the journey and seeking the support of appropriate guides along the way.

To return to the athlete metaphor, coaches can offer an external perspective, physiotherapists physical support and guidance, and training partners offer motivation to maintain consistent training. Each member of the support village plays a unique role; for example, no one would expect their coach to drag them through those early morning training sessions. Likewise, expecting a supervisor to provide all the support for a doctoral scholar is likely to cause frustration on both parts.

More than a support team, other peers within the support village can also act as valuable critical friends (Costa & Kallick, 1993). It is often through these critical friendships that learning of the hidden curriculum takes place. For example, in discussing a rejected funding bid or academic paper, conversations can be sparked about the practices and issues that underpin success in these areas. Having critical friends within a support village at a similar stage of their journey or a short way ahead on the road can also provide recent hot knowledge of these issues that academic and support staff may be one step removed from.

A Local Village for Local People

Each doctoral scholar is unique in terms of their discipline, personal and professional identities. This means each may have a slightly different idea of the 'perfect' village. Elliot et al. (2020) highlight the values of communities as playing a key role in doctoral thriving but for part-time doctoral scholars, the range of communities they are part of may be more complex and extensive. Often part-time doctoral scholars come to the doctorate with extensive personal and professional experiences and furthermore, these contexts maybe be nested and also contribute to doctoral success (McAlpine & Norton, 2006). Therefore, some of the elements of the hidden curriculum that are highlighted as 'hidden' to recent graduates entering a full-time doctorate may already be well understood. Likewise, some of the already understood elements by institutional graduates might appear alien. This creates an impetus underpinning the need to support individuals curate the village that they need in imperative.

The inhabitants of these villages, like any village, may change over time. Luckily with a support village, if a neighbour becomes a nuisance, it is easier to replace them than in an actual village. However, it can be hard for people to realise that. As such it is important to help doctoral scholars understand that people might, to use a cliché, come for a reason, a season, or a lifetime.

Institutional Scaffolding

If each individual needs to build their own village, then the role of the researcher developer should be to provide the space and support for doctoral scholars to understand who valuable additions to their village might be. The adage that you don't know what you don't know holds up here. In helping doctoral scholars to build their own villages, they need some candidates to be able to select from. For a doctoral scholar, the supervisor is a given, as might be a director of studies but there are innumerable support services that can also play a key role. Often institutions may signpost doctoral scholars to staff and services as crucial to doctoral success without ever explaining how they can help or why they are there. This can lead to them being sidestepped in favour of an overreliance on supervisors for information and support better provided by other services. A strong induction programme is likely to be a solution to some of this if it helps scholars to understand the why and how of the support available as opposed to simply providing a list of services.

Additionally, often the best people to promote their value for doctoral scholars are the individuals from those services themselves. Not only does this allow for the development of an interpersonal connection but they can also go beyond what their role involves and highlight how this can be of value at different stages of the doctorate. For example, knowledge of how invaluable librarians can be to research is often downplayed in a focus on the services a library offers.

Whilst Elliot et al. (2020) highlight the value of mentoring or coach-based approaches to support this, it is important to reflect on the time pressures of the part-time doctoral researchers who may not have the time or understand the value in their sort of training when they see the primary focus of their limited time as the thesis itself. They can therefore often see anything beyond working on key stages of the thesis with their supervisor as simply and additional drain on their time. This is not to argue that part-time scholars will not engage with these approaches but that three factors need to be carefully considered in the planning: accessibility, promotion of the value, and providing conditions for peer connections.

Accessibility
A prime consideration of course needs to be timetabling. Unlike full-time doctoral scholars, part-timers are likely to have a range of complex personal and or work commitments. Time to focus on the doctorate is often carved out between other things and as atypical times of the day or week. Therefore, activities focused within the working week are likely to be problematic. Equally when many part-time doctoral scholars are likely to be working at a distance, travelling to campus for a one-hour session is likely to be prohibitive. Often these barriers are not deliberate but created because researcher developers are not pausing to consider how to make these activities accessible.

What's in It for Them?
A secondary consideration is the often lack of an explicit rationale for why these activities matter. If informal interactions are essential to understanding the hidden curriculum, then this assertion itself needs to be clear and not hidden. For example, in Waterhouse et al. (2022) study of supporting professional doctorates at a distance, not all students saw the importance of interpersonal contact in their learning journey. This might be that those developing these activities have not spelled out the value of these activities and why they are a valuable investment of time, but it could also be that time stretch scholars are seeing the thesis as the necessary focus of their time and everything else as additionality and therefore not engaging with where learning about the hidden curriculum takes place.

Enabling Peer Connections
Beyond the initial support to identify who needs to be in support villages, some consideration needs to be made as to how to maintain the conditions for these connections to thrive. This is especially important in terms of providing the spaces and encouragement for peer networking. Peer learning builds accountability (Elliot et al., 2020, p. 70) and in a shared space, this is easier to encourage than at a distance. This therefore means rethinking how to encourage peer learning communities at a distance, although it can be done through online forums and virtual learning environments (Waterhouse et al., 2022) and Twitter (now *X*) (Vigurs, 2016).

Writing groups can also be an ideal space for learning about the hidden curriculum through experience sharing relating to writing practices but also persistence and self-motivation (Elliot et al., 2020, p. 87). Encouraging part-time doctoral scholars to join or run such groups can be invaluable.

These can also be a great opportunity for Early Career Researchers (ECRs) to support other doctoral scholars and pass on some of that hidden curriculum learning which simultaneously offering a space to focus on their own writing. (For a full discussion on doctoral writing groups, see Guerin and Aitchison's chapter.)

Developing Connections Beyond the Institution

It should not be assumed that all the inhabitants of a support village should come from within the institution. In fact, developing villages that span beyond institutional boundaries is likely to be invaluable in understanding elements of the hidden curriculum focused upon employment, the wider discipline, writing, and publication practices. Some ways researcher developers can help support doctoral scholars to expand their villages beyond the institution follow:

Social Media
Whilst Twitter and other social networks can be used to foster institutional communities, they also provide useful connections beyond the institution. Its immediacy, brevity, and ease of engagement can make it an invaluable tool for all academics (Carrigan, 2020). It can also provide a way to construct a professional academic identity (Sheldon & Sheppard, 2022) which can enable practical application of the learning from the hidden curriculum they are developing. Whilst it is important not to assume every doctoral scholar will want to engage with social media, for myself and many part-time doctoral scholars, it has provided an importance space for connection and reflection. They can offer a space for mutual support, accountability, and talking through ideas. However on the flip side they can be a time vortex, combative, or have a negative impact on confidence (Rainford, 2016; Rainford & Guccione, 2023). In supporting the use of these networks, it is important to forewarn of the potential issues that might be faced whilst also promoting the value of this for identifying and bringing in individuals to the support village who might be more geographically remote.

Postgraduate Subject Networks
From a researcher developer perspective, it is unlikely that we can be aware of every relevant subject association. Research projects are also rarely bounded in one neat discipline. For example, the topic of my own doctorate spanned several disciplines, each with their own learned societies:

Education, Sociology, Higher Education. Therefore, it is not as much about informing doctoral scholars of which networks to affiliate to but about providing the understanding of the importance of these networks, what the benefits might be, and then allowing them to find the networks that sustain them best. (For concrete examples of how research networks open the doors for research and learning opportunities, see Aarnikoivu's chapter.) This is, of course, something supervisors can help with but might require informal conversations with other researchers whose work sits within similar domains to the doctoral scholar, especially when the supervisor's own specialism is slightly different.

The Home-Village Connection

Whilst not directly supporting in learning the hidden curriculum per se, it would be impossible to talk about a support village without mentioning this key area in doctoral success. Home is being used in a broad sense to encompass friends, family, and work colleagues. Whilst some scholars might try to keep their doctoral and home life separate, there is likely to be an overlap and individuals from these parts of their lives are likely to be important to their journey. Practical and emotional support from these groups can be invaluable. In providing space for doctoral scholars to think about their home support villages, for some this may help identify hidden curriculum agents who have key skills that may help them—a friend or family member who is a great proof-reader, one who is a whizz with common software packages, or even an interested colleague who will provide the space to improve their communication skills to aid their oracy skills.

Conclusion

Returning to the initial metaphor of a hidden curriculum locked out of sight, the role of the researcher developer needs to go beyond unlocking the door and helping part-time doctoral scholars to find their own set of keys. Where full-time scholars might be able to ask someone nearby to borrow their key, the part-timers need access to their own set or a way to easily access them from someone else in their village. In doing so, it is more likely that as they navigate the maze of the doctorate and come up against the next locked door, they will be able to open that door with the help of someone in their support village and avoid being trapped in the unknown.

It is important for researcher developers to not just adopt a "build it and they will come approach" to this support though. It needs to be carefully planned and promoted so that it is accessible to these doctoral scholars at an appropriate time and that they understand the value for them in engaging with it. One valuable way to do this is by showing and not telling. Using existing doctoral scholars and postdoctoral researchers as examples of how their villages have been invaluable will make hesitant scholars more likely to engage than a list of 'selling points'. This also requires ongoing support and encouragement for informal spaces of support with consideration of how this can be done in both face-to-face and remote permutations.

Declarations of Conflicts of Interest None

Acknowledgements This chapter owes a debt to my own support village. Most importantly though, to the many part-time doctoral scholars myself and Kay Guccione have worked with over the past few years in developing our work on part-time doctorates.

References

Carrigan, M. (2020). *Social media for academics* (2nd ed.) Sage publications.

Costa, A. L., & Kallick, B. (1993). Through the lens of a critical friend. *Educational Leadership,* 51, 49–51. https://educandojuntos.cl/wp-content/uploads/2017/12/through_the_lens_of_a_critical_friend.pdf

Elliot, D. L., Bengtsen, S. S. E., Guccione, K., & Kobayashi, S. (2020). *The hidden curriculum in doctoral education*. Palgrave Macmillan.

Mantai, L. (2020). It takes a village to raise a PhD. *The hidden curriculum in doctoral education.* https://drhiddencurriculum.wordpress.com/2020/08/21/it-takes-a-village-to-raise-a-phd

McAlpine, L., & Norton, J. (2006). Reframing our approach to doctoral programs: An integrative framework for action and research. *Higher Education Research & Development, 25*(1), 3–17. https://doi.org/10.1080/07294360500453012

Rainford, J. (2016). Becoming a doctoral researcher in a digital world: Reflections on the role of Twitter for reflexivity and the internal conversation. *E-Learning and Digital Media, 13*(1-2), 99–105. https://doi.org/10.1177/2042753016672380

Rainford, J. (2021). Building your support tribe. *The hidden curriculum in doctoral education.* https://drhiddencurriculum.wordpress.com/2021/10/25/building-your-support-tribe

Rainford, J., & Guccione, K. (2023). *Thriving in part-time doctoral study: Integrating work, life and research.* Routledge.

Sheldon, J., & Sheppard, V. (2022). It started with a tweet: How doctoral researchers become social media savvy. In J. Sheldon & V. Sheppard (Eds.), *Online communities for doctoral researchers and their supervisors: Building engagement with social media.* Routledge.

Vigurs, K. (2016). Using Twitter to tackle peripherality?: Facilitating networked scholarship for part-time doctoral students within and beyond the university [Other Journal Article]. *Fusion Journal,* 8, 1–22. https://search.informit.org/doi/10.3316/informit.231887345934751

Waterhouse, P., Dennis, C. A., & Alvarez, I. (2022). Exploring perceptions of social presence among researching professionals. In M. G. Jamil & D. A. Morley (Eds.), *Agile learning environments amid disruption: Evaluating academic innovations in higher education during COVID-19.* Palgrave Macmillan.

PART IV

Insights from Institutional Leaders

Echo-locating a Personalised Route to Independence

Kay Guccione

Why Does Doctorate Need to Be Navigated?

A doctorate is a personal opportunity for learning about the self, as well as about the field and research topic of choice. Doctoral researchers have diverse past and ongoing lives, different personal and professional motivations for embarking on their doctoral journey, and a range of aspirations for their careers during and post-doctorate. It follows, then, that every doctoral researcher's learning experience will be different, and that different prior experiences, preferences, priorities, and time pressures will necessarily interplay to influence how they construct and navigate their own doctoral learning.

It can be a common, and hindering, misconception that once a doctorate has begun, there is a specific pathway to completion which will provide doctoral researchers with all the timely opportunities and resources they will need for success in an academic career. Operating on this assumption

K. Guccione (✉)
University of Glasgow, Glasgow, UK
e-mail: Kay.Guccione@glasgow.ac.uk

D. L. Elliot et al. (eds.), *Developing Researcher Independence Through the Hidden Curriculum*,
https://doi.org/10.1007/978-3-031-42875-3_16

can be the source of disorientation and frustration for doctoral researchers (McCulloch & Bastalich, 2023) and assuming that there is a single correct way to navigate through a doctorate can be a barrier to progress. Aligned with this, the benefits that individuals gain from their doctoral study are weighted and utilised in different ways by different people. Whilst one researcher may value the development of critical thinking skills as their most valuable doctoral asset, another may place highest value on their networks of friends and colleagues they have gained worldwide (Bryan & Guccione, 2018).

Within the dynamic intertwining ecology of doctoral education and development, there is potential for doctoral researchers to be overloaded with learning opportunities and to feel conflicted about what the most important and more urgent aspects of the doctoral learning experience are (Elliot et al., 2020, p. 47), creating feelings of confusion and uncertainty, which is experienced as discomfort (Albertyn & Bennett, 2021). When presented with the copious formal and informal learning opportunities that the formal and hidden curriculum offer, a honed sense of self-awareness supports doctoral researchers to locate and selectively engage with those opportunities that offer relevant and timely academic, personal, social, or psychological benefit.

This chapter conceptualises the acquiring of researcher independence within such a framework of self-awareness—as the gradual development of the ability to reveal, recognise, and evaluate one's own circumstances, experiences, and reactions, and to identify and act upon opportunities for learning, to achieve one's chosen career objectives (Elliot, 2022). Boud and Lee (2005) described this ability to learn to navigate the doctorate as becoming a 'self-organising agent' (p. 514) 'in which learners take up opportunities in a variety of ways without necessary involvement from teachers or supervisors' and demonstrated how researchers who are presented with similar opportunities understand and use the learning environment very differently, according to how they perceive themselves, their role, and their future (p. 503). Developing self-awareness and maturing to the point of becoming a 'self-organising agent' who does not require (but may still benefit greatly from) supervisor and teacher input involve an active reflective process (see also Elliot's discussion of the intentional use of metacognition).

However, for many it constitutes a hidden concept, one that is not discussed with supervisors, or features in study brochures. Becoming cognisant of our own values, preferences, attributes, and habits of thinking and acting is a gradual process, which is supported by feedback from and dialogue with others. For doctoral researchers, learning to evaluate how well their personal perspectives, behaviours, and habits are serving them in the pursuit of their study and career goals can be greatly enhanced by guiding conversations with an experienced mentor in the discipline, or with a near-peer a little way ahead. For a researcher who feels stuck or has lost momentum, and so is seeking to judiciously adjust their habits of thinking and acting in order to succeed in their endeavours, the support and encouragement of independent others can be transformational.

It is important therefore for doctoral researchers to know that self-awareness and reflective self-evaluation are important in their quest to develop independence (Elliot, 2022). Additionally, it is important for researchers to realise that these skills can be learned, developed, and embedded through regular practices. However, learning where, how, and with whom to engage in such transformative awareness-raising practices also commonly forms part of the hidden doctoral curriculum, increasing the complexity of accessing them.

Transformation Through Conversation

Self-awareness, self-evaluation, and prioritising what is relevant and required to transition towards independence can be supported through opportunities to engage in critical reflective discourse and is situated within the theory of Transformative Learning (Mezirow, 1991). Transformative Learning theory focuses on the idea that learners have the capacity to adjust their established thinking patterns and ways of ascribing meaning to new events, based on the critical evaluation and reflective embedding of that new information. Having understanding of our established perspectives and habits, and the ability to unpick these to weave in new ideas and perspectives, is key to Transformative Learning theory. As such, Transformative Learning describes a mechanism for awareness raising, self-evaluation, and, in light of this, forward planning.

Mezirow's theory states that Transformative Learning is initiated as a 'Disorienting Dilemma'. Many scholars will recognise this as being a prominent and recognisable phenomenon in the doctorate, as old ways of approaching learning and generating academic success prove to be ineffective within the doctorate's more open and individualised learning framework. Following the model, a Disorienting Dilemma leads to a phase of self-examination, and then to a critical assessment of prior assumptions, before planning a new course of action. Proponents of Transformative Learning theory (notably established through classroom-based observations of adult learners) suggest that this process can be enhanced by creating structures for reflective conversation, preparing relevant literature and reflective activities, and giving students opportunities for dialogue and debate. Within the doctoral education model, such opportunities to critically reflect on the personal impact of new learning experiences are not always so obvious. So, where, and how, can we create the conditions to ensure that the inevitable doctoral Disorienting Dilemmas candidates will experience can be productively resolved into transformative action?

Drawing on the practices of a coaching conversation, involving exploratory, reflective, and non-judgemental dialogue, a range of different agents of the hidden curriculum can support doctoral learners to separate out their complex and intertwined thoughts, and to resolve instances of disorientation that have arisen. Acting as a 'sounding board'—that is, 'seeking to amplify the researcher's own voice' (Guccione & Hutchinson, 2021, p. 65)—a range of hidden curriculum agents in varied roles can provide opportunities for reflective dialogue. Skilled coaching, mentoring, and peer and supervisor conversations can all support researchers to develop their own abilities to self-evaluate and to devise plans for seeking and engaging with additional learning experiences. In doing so, they support the developing doctoral candidate to vocalise their thoughts, feelings, and ideas out loud, enabling them to build self-awareness, test ideas, and 'echo-locate' their own individual pathway towards degree completion and therefore towards independence.

Creating an intellectually and emotionally personalised learning plan, the result of the purposeful partitioning of time to engage in rational dialogue, has the additional advantage of bolstering researchers' perceptions of certainty and control over their learning and learning environment. Gaining clarity about what has been learned and how it can be useful is often experienced as feelings of reassurance, or even relief, as uncertainties are resolved (Albertyn & Bennett, 2021). A conversational 'sense check' or 'validation of plans' can also support doctoral researchers to gain the confidence to adapt their study strategies to move in a positive direction (Godskesen & Kobayashi, 2016). What is more, an intentionally caring conversation covering both the intellectual and emotional challenges of doctoral work can work to enhance the quality of supervisory relationships (Caretta & Faria, 2020) and support wellbeing (Griffin et al., 2023). Intentional transformative dialogue has been deployed as an effective tool in the decolonisation of doctoral, academic, and higher education spaces (Rispel, 2023).

Given the potential range of direct and indirect benefits of reflective dialogue above, the final part of this chapter offers some ideas for practice.

Designing Discourse into the Doctorate

There are, of course, many opportunities for supportive conversations within the doctorate. However, what I advocate for here is a purposefully designed awareness-building conversation with specified rules of engagement, and that requires the use of specific skills by the supporting interlocutor. We begin by considering how to create the conditions for a conversation where the primary aim is to raise self-awareness, to offer opportunities for self-evaluation, and thus to resolve feelings of disorientation. The role of the supporting interlocutor (be they supervisor, disciplinary peer, or development professional) is to support the doctoral researcher to think out loud, providing the time, space, and permission to think and to hear themselves think.

To create the right conditions for this, we must be aware that the goal of the conversation is not to impart the wisdom of our own experience or to give advice, but to listen and to pay attention to what is said. In doing so we support the shift into a state of independent decision making, away from reliance on the knowledge, skills, and experiences of a more senior person. Consider that in such a conversation, talking about oneself (*our* advice, *our* opinion, *our* experience) is antithetic to the idea of building independence, as it focuses the conversation away from the doctoral researcher. Additionally, being interrupted with well-meaning advice, anecdotes, or related topics can be experienced as frustrating, invalidating, and disempowering by that researcher.

This style of reflective conversation is challenging work, and it takes practice, as it requires all supporting interlocutors to put aside their own agendas, pressures, and opinions, along with any ready-made solutions they feel inclined to offer. More so as the giving of advice and recommendations, and the sharing of expertise, is affirming within our identities as academic experts. We must de-centre ourselves. However, the supporting interlocutor role is not a passive one. We can offer input and insight and create structure to frame the conversation, but we must listen very carefully to take our cues from the doctoral researcher's experience, rather than our own. We can utilise the ideas adapted from Guccione and

Hutchinson's (Guccione & Hutchinson, 2021, p. 70) work in this area. Within the conversation a good supporting interlocutor will:

- **Support self-awareness.** Encourage reflection on the thoughts and feelings that are experienced, the behaviours and actions displayed, and any assumptions, values, or perspectives that led to those.
- **Focus attention.** Recognise when a number of intersecting issues have been mentioned inviting the doctoral researcher to focus into each element of the issue, choosing which is the most pressing priority.
- **Support self-evaluation.** Ask questions such as 'What prevented that from going how you planned it?' 'What can we learn about what gets in the way of your progress?' This helps the learner to focus on recognition of tangible obstacles that can be defined and overcome.
- **Recap the conversation.** Repeating back on what you have heard can be a powerful sense-making tool. To summarise is to recap what you have heard in your mentee's own words. To paraphrase is to give a short precis using your own phrasing of the situation. Both can prompt reflection.
- **Listen beyond the words.** Listen for excitement or disappointment, be sensitive to their changing energy levels, and notice what parts of their work they do not seem enthusiastic or energetic about. Feed this back into the discussion.
- **Support small actions.** What are the next steps for the researcher? How can large aims be segmented into smaller objectives? Which needs to be done first? Support identification and reservation of appropriate time to complete defined small goals.

With these conversational tools at hand, opportunities to support the development of independence through reflective dialogue can be made and strengthened in many ways within the design of doctoral programmes. To choose a few examples, the above style of reflective dialogue can be built into:

- Induction workshops, cohort building activities, and expectation setting with supervisors.
- Setting up buddying or peer-mentoring partnerships.
- Every supervisory meeting.

- Support for formal processes such as Development Needs Analyses and Annual Progress Reports.
- Preparation for and debriefing of doctoral milestones such as conference presentations, or peer review.
- A research team approach to regular progress meetings that avoids reliance on a 'hub and spokes' model of supervision.
- Community-building activities in departments and schools.
- Career conversations, job applications, and interview preparation.
- Valuing and prioritising discursive spaces within taught classes, training, and workshops of all kinds.

In conclusion, this chapter positions researcher independence as the ability to recognise and navigate one's own circumstances and experiences when presented with copious formal and informal learning opportunities, and to critically select and engage with those that offer academic, personal, social, or psychological benefit. Self-awareness and the skills of self-evaluation can be developed through the Transformative Learning model which prioritises dialogic modes of reflection (see also Elliot, 2022). Each reflective conversation represents an access point for learning from the hidden curriculum and for the gaining of independence as a developing researcher.

References

Albertyn, R., & Bennett, K. (2021). Containing and harnessing uncertainty during postgraduate research supervision. *Higher Education Research & Development, 40*(4), 661–675. https://doi.org/10.1080/07294360.2020.1775559

Boud, D., & Lee, A. (2005). 'Peer learning' as pedagogic discourse for research education. *Studies in Higher Education, 30*(5), 501–516. https://doi.org/10.1080/03075070500249138

Bryan, B., & Guccione, K. (2018). Was it worth it? A qualitative exploration of graduate perceptions of doctoral value. *Higher Education Research and Development, 37*(6), 1124–1140. https://doi.org/10.1080/07294360.2018.1479378

Caretta, M. A., & Faria, C. V. (2020). Time and care in the "lab" and the "field": Slow mentoring and feminist research in geography. *Geographical Review, 110*(1–2), 172–182. https://doi.org/10.1111/gere.12369

Elliot, D. L. (2022). A 'doctoral compass': Strategic reflection, self-assessment and recalibration for navigating the 'twin' doctoral journey. *Studies in Higher Education, 47*(8), 1652–1665. https://doi.org/10.1080/03075079.2021.1946033

Elliot, D. L., Bengsten, S. S. E., Guccione, K., & Kobayashi, S. (2020). *The hidden curriculum in doctoral education*. Palgrave Macmillan. ISBN 9783030414962; https://doi.org/10.1007/978-3-030-41497-9

Godskesen, M., & Kobayashi, S. (2016). Coaching doctoral students—A means to enhance progress and support self-organisation in doctoral education. *Studies in Continuing Education, 38*(2), 145–161. https://doi.org/10.1080/0158037X.2015.1055464

Griffin, K. A., Stone, J., Dissassa, D.-T., Hall, T. N., & Hixson, A. (2023). Surviving or flourishing: How relationships with principal investigators influence science graduate students' wellness. *Studies in Graduate and Postdoctoral Education, 14*(1), 47–62. https://doi.org/10.1108/SGPE-12-2021-0085

Guccione, K., & Hutchinson, S. (2021) *Coaching and mentoring for academic development*. Series: Surviving and Thriving in Academia. Emerald Publishing Limited. ISBN 9781789739107

McCulloch, A., & Bastalich, W. (2023). Commencing research students' expectations and the design of doctoral induction: Introducing inflections of collaboration and pleasure. *Journal of Further and Higher Education, 47*(5), 687–698. https://doi.org/10.1080/0309877X.2023.2185772

Mezirow, J. (1991). *Transformative dimensions of adult learning*. San Francisco: Jossey-Bass Inc Pub.

Rispel, L. C. (2023). Relationships, power and accountability: Reflections on transformative postgraduate supervision workshops at a South African University School of Public Health, 2014–2020. *Innovations in Education and Teaching International, 60*(1), 101–111. https://doi.org/10.1080/14703297.2021.1952888

Decolonising Doctoral Education: Sociology of Emergences?

Catherine Manathunga

Introduction

The COVID-19 pandemic, the Black Lives Matter movement, and the climate change crisis have created greater urgency around the need to decolonise all forms of education, including doctoral education. Global doctoral education has been significantly affected by the COVID-19 pandemic. Public health risks and international border closures since early 2020 have caused disruptions, delays, and adaptations for international doctoral scholars. Progress of doctoral scholars in the global South has had to be deferred as people struggle to remain healthy, retain their employment, and care for those in their family that are sick or unemployed (Bob et al., 2021). The death of George Floyd in the United States and subsequent Black Lives Matter (BLM) protests around the globe have very powerfully drawn world attention to the need for decolonisation across a range of institutions.

C. Manathunga (✉)
University of the Sunshine Coast, Maroochydore, QLD, Australia
e-mail: cmanathu@usc.edu.au

D. L. Elliot et al. (eds.), *Developing Researcher Independence Through the Hidden Curriculum*,
https://doi.org/10.1007/978-3-031-42875-3_17

Internationally, we are at a crucial tipping point in the fight against all forms of racism and unconscious bias. The climate change crisis, which majority of world nations have been experiencing for many years, has also now become evident all around the globe with unprecedented floods, fires, droughts, and other natural catastrophes. We can no longer ignore the dire warnings of scientists and, as always, it is the poorest communities that disproportionately suffer the effects of climate change. All of these global trends ensure that there has never been a more urgent time to decolonise doctoral education.

Doctoral education, as a key site of knowledge creation and the preparation of future generations of scholars, occupies a unique location for institutional leaders, academics, and doctoral scholars to address these global challenges and their disproportionate effect on First Nations, Southern, and transcultural communities. Doctoral education, especially in systems where doctoral scholars work intensively on an individual research project with little or no formal coursework under the guidance of two or more supervisors, is often made up of a highly individualised and implicit curriculum. Just as important as any formal learning about research are the forms of informal learning that Elliot et al. (2020) describe as the hidden curriculum in doctoral education.

All doctoral scholars are able to access the hidden curriculum if they are provided with support to recognise learning opportunities in a vast array of settings, locations, and from a wide selection of people. These kinds of 'academic, emotional, social and psychological support[s]' are vital for all doctoral scholars but especially for international scholars 'as they settle into a "foreign" learning culture and environment' (Elliot et al., 2020, p. 14). I would argue they are also particularly useful for other transcultural (migrant, refugee, culturally diverse and international students) and non-traditional (Engels-Schwarzpaul & Peters, 2013) (e.g. working class or first in family) scholars who may be unfamiliar with research learning cultures. Also, First Nations, Southern, transcultural, and other non-traditional scholars bring with them often hidden or unacknowledged reservoirs of knowledge, agency, and networks that can provide important support structures for doctoral scholars.

These additional resources may remain opaque or invisible if doctoral education continues to be framed in ways that only acknowledge Northern or Eurocentric knowledge systems. If we decolonise doctoral education around the globe, we will be able to create spaces where First Nations, Southern, and transcultural doctoral scholars' knowledge, skills, connections, and communities are valued and where we may learn more about how best to support all of our doctoral scholars.

Therefore, it is important for institutional leaders, supervisors, and doctoral scholars to develop better recognition of these hidden 'doctoral treasure[s]' (Elliot et al., 2020, p. 4) that First Nations, Southern, and transcultural doctoral scholars possess. In this chapter, I explore how these features of the hidden doctoral curriculum can be reconceptualised as examples of Boaventura de Sousa Santos' (2018) concept of the 'sociology of emergences'. The sociology of emergences are ideas, philosophies, and practices that are centred upon what de Sousa Santos (2014) calls the 'epistemologies of the South' or Indigenous, Southern, and transcultural knowledge systems. These sociologies of emergences are presently part of the hidden curriculum of doctoral education.

In this chapter, I will share some practical examples from Australia and Aotearoa New Zealand about how these sociology of emergences build agency and researcher interdependence especially among First Nations and transcultural doctoral scholars. I am writing this chapter as an Irish-Australian settler-invader academic with a transcultural family, who has researched doctoral education and supervised doctoral scholars for around three decades, particularly in Australia and Aotearoa New Zealand. I am currently the Deputy Head of School (Research) of the School of Education and Tertiary Access at a small regional university in Queensland, Australia, and work with our Higher Degrees Research Coordinator to lead doctoral education in Education.

Sociology of Emergences and the Hidden Curriculum

De Sousa Santos (2018) argues that there are three forms of sociology of emergences contained within the epistemologies of the South. These include 'ruin seeds, counterhegemonic appropriations, and liberated zones' (de Sousa Santos, 2018, p. 29). Ruin seeds are the 'absent present' echoes of precolonial histories that provide ideas for a future beyond capitalism, colonialism, and patriarchy (de Sousa Santos 2018, p. 29). An example of ruin seeds is the African philosophy of Ubuntu (de Sousa Santos, 2018), which comes from several African languages, especially the Nguni languages of Zulu and Xhosa in South Africa. Ubuntu was described by Nelson Mandela as 'the profound sense that we are human only through the humanity of others; that if we are to accomplish anything in this world, it will in equal measure be due to the work and achievements of others' (Mandela, 2009, p. ii). Ubuntu also encapsulates the *inter*dependence doctoral scholars need to develop during their research studies to become highly accomplished future researchers.

Counterhegemonic appropriations are theories, concepts, and strategies used by dominant groups but then subverted and altered by oppressed groups to challenge the status quo (de Sousa Santos, 2018). De Sousa Santos says an example of counterhegemonic appropriations includes ideas where the Eurocentric concept of liberal democracy is reimagined to reformulate economic relations (post-capitalist), transcultural relations (post-colonialism), and gender relations (post-patriarchy) (de Sousa Santos, 2018). Liberated zones are ideas and approaches constructed according to Southern knowledge systems where 'principles and rules [are] radically opposed to those that prevail in capitalist, colonialist and patriarchal societies' (de Sousa Santos, 2018, p. 31). Liberated zones are also those where people 'live today as if today where the future to which they aspire' (de Sousa Santos, 2018, p. 31). This has also been described as 'prefigurative politics' (Raekstad & Gradin, 2020). An example of liberated zones includes the neo-Zapatista Indigenous communities in southern Mexico who, after 1994, began to form their own self-governing institutions, education and justice systems, healthcare, and agrarian and economic relations because their attempts to negotiate change with the State had failed (de Sousa Santos, 2018).

If we reformulate the concept of the hidden or informal curriculum in doctoral education through the lens of de Sousa Santos' (2018) ideas about sociology of emergences, then we could think of the vast array of social and cultural networks First Nations and transcultural doctoral scholars bring with them into doctoral education as the 'ruin seeds' or the echoes of precolonial ideas that can be drawn upon to create different futures in doctoral education. In many First Nations and transcultural cultures, philosophies of reciprocity, relationality, collectivism, interdependence, and community are central. Rather than an individualised and highly abstract focus, First Nations and transcultural doctoral scholars often bring with them a desire to use their research to solve practical community problems. These networks also hold vast stores of knowledge, wisdom, and philosophical, theoretical, and practical resources. In the following sections, I will explore some of these cultural resources and networks within academe and beyond universities by tracing examples where

Indigenous and transcultural cultural resources, networks, Elders, traditional knowledge holders, and intergenerational communities have been used in supporting doctoral scholars.

First Nations and transcultural knowledge systems are vital for the future of our planet and contain both ruin seeds of ancient wisdom refracted across time and space into dynamic and generative insights for the future as well as counter-hegemonic appropriations. First Nations and transcultural doctoral scholars are able to recast Western/Northern concepts, theories, ideas, and practices into unique new strategies designed to encourage the reconfiguring of economic, transcultural, and gender relations. In the following sections, I will provide some concrete examples of situations where Indigenous and transcultural knowledge systems have been used to reconceptualise Northern concepts in research.

Finally, I will apply de Sousa Santos' (2018) idea of liberated zones to recommendations for doctoral institutional leaders that might try to create the kind of future we desire where First Nations and transcultural knowledge systems, agency, and networks are fully recognised. These kinds of liberated futures or prefigurative politics would not only benefit First Nations and transcultural doctoral scholars' but may also help institutional leaders to better support all doctoral scholars.

There is also one other crucial area of support that current research on the hidden curriculum has not yet foregrounded and that is the significance of *spiritual* support for doctoral scholars. Often conceived of in highly secularised, political spaces of thought, the hidden doctoral curriculum refers to 'academic, emotional, social and psychological support[s]' (Elliot et al., 2020, p. 14). I would argue that all doctoral scholars may also require some form of spiritual support. This can take many forms, but First Nations and transcultural knowledge systems are often built around understandings of the role of spirituality in being human, connecting with our more-than-human environments and doing research. I, therefore, include a section below that explores, through examples (for more detail see Manathunga, 2020), the role of ancestors and spirituality in the hidden reservoir that Indigenous and transcultural doctoral scholars may bring with them into their research. This section is included in the section on ruin seeds and adds to my recommendations for creating a liberated zone of decolonised doctoral education.

'Ruin Seeds' Supporting the Hidden Doctoral Curriculum: Cultural Resources, Elders, and Communities

Indigenous and transcultural doctoral scholars are now being encouraged to use the concepts, proverbs, and wisdom from their own languages and cultural knowledge systems in their doctoral research. These cultural resources act as a form of 'ruin seeds' or ancient understandings repurposed in the present for new forms of contemporary knowledge. My research collaborator, Qi Jing, a Chinese-Tibetan-Mongol woman, who studied and now works in Australia, produced a doctoral thesis on transnational early childhood education in Chile. She used Chinese metaphors and language to 'critique transnational education and develop new pedagogical approaches' (Qi, 2015, p. 194). One of the Chinese metaphors she used was the idea of a 'networked-hutong siwei' (Qi, 2015, p. 37). Hutongs are the narrow back alleyways that are characteristic of residential areas in Northern Chinese cities like Beijing. Many of these hutongs are interconnected and labyrinth-like. *Siwei* is the Chinese word for thinking. Adopting a networked-hutong siwei in transnational education, Jing argues, allows educators to explore multiple approaches and to work with diversity as a creative strength built upon assumptions of intellectual equality and respect. (See also Rainford's chapter on harnessing strengths from support villages for intellectual thriving and emotional and practical support.)

A vital source of support for Indigenous and transcultural doctoral scholars includes special networks including Elders, traditional knowledge holders, and intergenerational community members. For example, in Aotearoa New Zealand and Australia these networks are involved in doctoral programs as co-supervisors and sometimes as examiners. Grant and McKinley (2011) outline the significant roles played by supervisors external to the university. These include the doctoral scholars' ***kaumātua*** (Māori male or female Elders) who take on a research-related, grandparent-type relationship with them. First Nations Australian Linda Ford (2012) also had supervisors who were community Elders.

Tracey Bunda and our transcultural research team acknowledged the Arrernte Community as co-authors on a book chapter we wrote where we drew upon Arrernte language and concepts to talk about doctoral education to demonstrate how Indigenous knowledge is collectively owned and shared (Manathunga et al., 2020). Tracey's adopted daughter, Angie, is an Arrernte woman from the Alice Springs area in Australia's Northern Territory.

Extending the Hidden Doctoral Curriculum to Include Spiritual Knowledge

One aspect of the hidden doctoral curriculum that has not yet been acknowledged is the importance of ancestors and spirituality in doctoral research. In a number of documented cases, Indigenous and transcultural doctoral scholars have sought to access ancestral knowledge through ritual, dreams, and spiritual practices. Grant and McKinley (2011) describe the role played by Māori doctoral scholars' *tūpuna* (female or male ancestor), who provide spiritual guidance for their research. Devos and Somerville (2012) describe working with a Cambodian doctoral scholar in Australia. The student had been given a copy of the memoir of her grandmother who was from the Cambodian Royal Family. She was forced to rely only on her memory to transcribe and translate this memoir into English in writing her thesis because her family members had been killed—and the original text destroyed—during the Pol Pot regime. The student believes that her grandmother guided her towards meeting her Australian supervisors (Devos & Somerville 2012).

Ruins Seeds and Counterhegemonic Appropriations: First Nations and Transcultural Knowledge Systems

Contemporary Indigenous and transcultural knowledge systems contain both ancient wisdom and counterhegemonic appropriations of Western/Northern concepts. For example, *Kaupapa Māori* is a well-established theoretical paradigm used in Aotearoa New Zealand. Literally, *kaupapa* means ground rules (Smith & Reid, 2000). *Kaupapa Māori* is the 'systematic organisation of beliefs, experiences, understandings and interpretations of the interactions of Māori people upon Māori people and Māori people upon their world' (Nepe, 1991, p. 76).

In Australia, there are a range of theories, ontologies, and methodologies embedded in the notion of Indigenous Knowledge (IK) which are used across the sciences, engineering, health, agriculture, education, arts, humanities, and the social sciences. Aboriginal academic Norm Sheehan (2011, p. 69) states that

> IK operates from the assumption that the world is alive and active in the same way that humans are alive and active. Respect is based on this ancestral understanding that we all stand for a short time in a world that lived long before us and will live for others long after we have passed. From this view, we can never know the full implications of any action; thus, IK respect is about showing care and awareness in the way we identify, explore, and assess meaning because we know our view is always incomplete.

Indigenous Knowledge research methodologies, such as the yarning circle, are now frequently employed in Australia. As Sheehan (2011, p. 70) outlines, 'yarning circles are conducted under the simple rules that each person speaks in turn, holds authority for the time they speak, and reciprocates by speaking responsibly from self and not about others'.

Conclusion: Liberated Zones for Decolonised Doctoral Education—Ideas for Doctoral Institutional Leaders

What would a liberated zone of decolonised doctoral education look like at this moment in the twenty-first century? How might such a liberated, decolonised idea of doctoral education acknowledge the many learning opportunities that exist in hidden, informal settings with a range of people? If you are an institutional doctoral leader, you will need courage and the desire to swim against the neoliberal tsunami churning over doctoral education at present to design and implement liberated zones in doctoral education. Liberated zones are about making the choice to act as if the future forms of doctoral education we seek, like decolonised doctoral programs, are already here.

This would involve acknowledging and actively encouraging Indigenous and transcultural doctoral scholars to draw upon their own historical, geographic, cultural, epistemic, spiritual, and linguistic resources, networks, Elders, traditional knowledge holders, and intergenerational communities as vital supports for their doctoral research. These strategies would emphasise the need for students to develop interdependence (or a sense of Ubuntu) in their approach to research that recognises the important role families, communities, and collectivities play in all stages of the research process. Institutional leaders could (and in some cases already do) formalise the significant role of First Nations and transcultural Elders, traditional knowledge holders, and intergenerational communities as co-supervisors or, where appropriate, examiners of doctoral research.

It would also include understanding and promoting the importance of spiritual support in doctoral education as an additional feature of the hidden doctoral curriculum. Leading change around spiritual support in doctoral education would involve leaders ensuring that guidelines acknowledge doctoral scholars may require spiritual as well as emotional and cultural support during their studies and provide a list of possible contacts that could be called upon. These contacts could include academics, Elders, and traditional knowledge holders attached to Indigenous support units, chaplains, university counselling services, or other spiritual community groups. All of these important forms of academic, emotional, social, psychological, *and* cultural and spiritual support can be seen as the ruin seeds that de Sousa Santos suggests are features of a decolonised sociology of emergences.

Creating liberated zones within doctoral education would also mean that institutional leaders would need to fully acknowledge, recognise, and accredit Indigenous and transcultural knowledge systems that contain the ruin seeds of ancient, precolonial, and contemporary First Nations and transcultural wisdom, as well as counterhegemonic appropriations of Western/Northern concepts. It would involve encouraging Indigenous and transcultural doctoral scholars to use First Nations and transcultural philosophies, theories, ontologies, and methodologies in their research. Institutional leaders could also act as key advocates for the validity and significance of Indigenous and transcultural knowledges and the importance of developing interdependence in research. In these ways, we might apply de Sousa Santos' concept of the sociology of emergences in order to begin working towards a decolonised form of doctoral education.

Declarations of Conflicts of Interest None

Acknowledgements I would like to acknowledge my colleagues Jing Qi, Maria Raciti, Michael Singh, Tracey Bunda, Kathryn Gilbey, Dr Aunty Hope O'Chin, Dr Aunty Sue Stanton and Dr Aunty Judi Wickes for teaching me so much about decolonisation. I would also like to acknowledge Barbara Grant and Avril Bell for inspiring conversations during the writing of this piece that helped sharpen my thinking.

References

Bob, U., Munien, S., Gumede, A., & Gounden, R. (2021). Postgraduate training and supervision and post the COVID-19 crisis in South Africa. In T. Mgutshini, K. Oparinde, & V. Govender (Eds.), *COVID-19: interdisciplinary explorations of impacts on higher education* (pp. 131–145). African Sun Media.

De Sousa Santos, B. (2014). *Epistemologies of the South: Justice against epistemicide.* Paradigm Publishers.

De Sousa Santos, B. (2018). *The end of the cognitive empire: the coming of age of epistemologies of the south.* Duke University Press.

Devos, A., & Somerville, M. (2012). What constitutes doctoral knowledge? *Australian Universities Review, 54*(1), 47–54.

Elliot, D., Bengtsen, S., Guccione, K., & Kobayashi, S. (2020). *The hidden curriculum in doctoral education.* Palgrave Macmillan.

Engels-Schwarzpaul, A., & Peters, M. (Eds.). (2013). *Of other thoughts: Non-traditional ways to the doctorate.* Sense Publishers.

Ford, L. (2012). Reflecting on doctoral candidature experiences as an Indigenous Australian. In C. Denholm & T. Evans (Eds.), *Doctorates downunder* (pp. 145–152). ACER Press.

Grant, B., & McKinley, E. (2011). Colouring the pedagogy of doctoral supervision: Considering supervisor, student and knowledge through the lens of indigeneity. *Innovations in Education and Teaching International., 48*(4), 365–374.

Manathunga, C. (2020). Decolonising higher education: Creating space for Southern sociologies of emergence. *SoTL In the South, 4*(1), 4–25.

Manathunga, C., Bunda, T., Singh, M., Qi, J., & Community, T. A. (2020). Indigenising the higher degree research space: Ampere tyerrtye-areye apurte-irrentye akngerre. Tyerrty-arey-ae antartaremele mwarre anetyeke (A place where people come together and look after each other). In K. Trimmer, D. Hoven, & P. Keskitalo (Eds.), *Indigenous postgraduate education: Intercultural perspectives* (pp. 89–104). IAP.

Mandela 'Preface' in Stengel, R. 2009. *Mandela's Way: Fifteen Lessons on Life, Love, and Courage.*

Nepe, T. (1991). *E hao nei e tenei reanga te toi huarewa tupuna: Kaupapa Māori, an educational intervention system.* Unpublished MA Thesis, Education Department, University of Auckland, Australia.

Qi, J. (2015). *Knowledge hierarchies in transnational education: staging dissensus.* Routledge.

Raekstad, P., & Gradin, S. (2020). *Prefigurative politics.* Polity Press.

Sheehan, N. (2011). Indigenous knowledge and respectful design: An evidence-based approach. *Design Issues, 27*(4), 68–80.

Smith, L. T., & Reid, P. (2000). *Māori research development: Kaupapa Māori principles and practices—A literature review.* Te Puni Kokiri.

Midwifing the New: Institutional Leadership for Doctoral Education

Barbara M. Grant

An Intriguing Triangle for Thought

Juxtaposing ideas of leadership for doctoral education, hidden curriculum and researcher independence creates an intriguing triangle for thought. Each of these ideas (fruitfully) troubles me in some way. First, leadership because of the way it has been captured in universities by a hierarchical management structure of command and obey alongside a culture of widespread 'leaderism' that privileges the "power of personality" and "individual agency" (Macfarlane, 2014, p. 2). Then, hidden (or informal) curriculum because, along with Bill Green (2018), I'm unsure what the *formal* curriculum of doctoral education actually is, let alone how we

B. M. Grant (✉)
Waipapa Taumata Rau/University of Auckland, Auckland, New Zealand
e-mail: bm.grant@auckland.ac.nz

D. L. Elliot et al. (eds.), *Developing Researcher Independence Through the Hidden Curriculum*,
https://doi.org/10.1007/978-3-031-42875-3_18

might think about its hidden counterpart.[1] And, lastly, researcher independence because I find thinking about researcher *inter*dependence more compelling, although I understand independence as implicated there.

In thinking about each of these ideas more fully, I draw on varied institutional experience as a member of the university's governing board for doctoral education, provider of professional development for supervisors, school doctoral advisor and supervisor. I begin by sketching the landscape of institutional leadership for doctoral education and sharing some grief over its contours. The second part of my chapter takes a more hopeful turn as I sketch possibilities for contributing to the formation of doctoral scholars in more local and embedded spheres of leadership. There we can initiate practices designed to foster an appreciation of the interdependence required to flourish as thinkers, readers, researchers and writers. This involves learning about the ethical obligations that interdependence entails, including that of *giving time* to others. The figure that arises during my consideration of this kind of leadership, inspired by Green's (2018) reference to natality, is that of the midwife (see Haynes, 2009, for an exploration of midwifing as a metaphor for both doctoral work and supervision). In doctoral education, leading by midwifing is the role of helping others arrive in the world—others who, in turn, are capable of the interdependent labour involved in creating not only new knowledge but also new academic subjects.

Institutional Leadership Is Compromised

In my university, as in many others, institutional leadership of doctoral education at the top level is relatively recent. Not so long ago, there was just a statute and a detailed guide to compiling a thesis (concerned with page margin widths and the order of front material, etc.). These documents, along with final examination decisions, were overseen by a Board of Graduate Studies with representation from every faculty as well as select

[1] I have previously suggested that, in the absence of courses, the doctoral curriculum must be continually inferred from diverse sources such as supervisory guidance, provisional year goals, generic doctoral skills workshops, doctoral graduate profiles, and examination criteria (Grant, 2011). Proposal writing and completed theses are helpful too. What I described as an 'inchoate' curriculum, Green (2018) explores as an 'emergent curriculum': "What is produced in the doctoral project, in active, open dialogue with the archive and the world? Curriculum in this context is something that emerges and is realized, only retrospectively, in a new articulation of knowledge and identity" (p. 79).

other interested parties (such as careers services and postgraduate students). Since the early 2000s, however, we have had both a Dean and a School of Graduate Studies. Initially, the Dean and School oversaw the administrative aspects of doctoral education only. Over time, though, the Dean's role has led to a whole new level of quality- and performance-driven activity in relation to doctoral education: mandatory supervisor[2] induction and 'training'; a mandatory induction day for new doctoral students and expanded provision for academic skill development; ambitious student recruitment targets; vastly increased banks of documentation (policies and forms); frequent revisions to the statute; regular institution-wide (and comparative between faculties) surveillance of enrolments, progress, completions, terminations and examination outcomes; international forays to recruit new students; and so on. Most recently, the Dean has enthusiastically endorsed a new online system for handling all the administrative aspects of a student's candidature which, under the guise of efficiency, has radically enhanced the potential for surveillance of supervisions. In this climate, policy compliance is paramount and dispiriting: "It's easier to get a mortgage for a house than to fill out a form for your doctoral student" (Alison Phipps, PaTHES webinar conversation, 07/10/22).

A problem for top-level doctoral leadership is that, like other high-level university leadership, it is caught in competing demands from single-interest 'stakeholders', the toxic mix of commodification and boundless opportunity characteristic of academic capitalism (Slaughter & Leslie, 2001), and many forms of international benchmarking and ranking that lead to considerable standardisation of practice. In addition, senior university leadership is disarmed by a culture of high rewards (usually via independent employment contracts) in return for consummate obedience to whoever sits above them. In doctoral education, there are specific concerns about long-standing patterns of slow completion and attrition of doctoral students and how supervision might be implicated. There are also leaner government funding regimes and high levels of sensitivity to both risk for institutional reputation *and* advantage from public research funding allocation and international rankings (doctoral enrolments and completions count for both).

In my experience, there are two common top-level leadership responses to this pressure: First, doctoral education is increasingly policed, drawing heavily on norms derived from STEM models and embedding them into

[2] Also called (especially in the US) 'advisors'.

generic institutional policies. Second, a transactional relationship between customer-student and provider-university is reinforced over and over. Many of those transactions are frustrating to enact, and sometimes actively subverted by supervisors (and students), as a recent PhD Director describes:

> *This institutional form dates from when I was the director of our PhD programme. I was frustrated at how some colleagues passively-aggressively refused to use the forms in ways that served their students, and often acted as if the university had no right to ask for accountability. I'm not the director now. I often wish there was a job where I could do more developmental work with PhD students and supervisors, rather than the form-filling role.* (Virginia, Retreat 2016, Workshop 1)[3]

Such a confluence of actions and reactions produces a not-so-hidden curriculum for supervisors and students alike, in which supervisor and student compliance and student satisfaction are premium goods. Yet what is sought from an academic perspective (originality and independence) is something quite different. In such a context, re-imagining top-level leadership towards 'fostering' any meaningful domain for doctoral education, especially possibilities for cultivating student interdependence, is difficult, as Virginia remarks here.

Cascading Layers of Leadership

To describe my (by no means unique) university again, the position of the Dean of Graduate Studies cascades: there are doctoral education leadership roles at the level of faculty (or college) and then school (or department). Because of the command and obey model of leadership in the university more widely, the faculty-level position (typically called Associate-Dean Postgraduate) often simply propagates decisions coming from above. Moreover, encouraged by the competitive culture of the academy, these mid-level position-holders sometimes seek to excel in enacting them. Virginia and Donna, two participants in my study of women doctoral supervisors in Aotearoa New Zealand, talked about their experiences of local (school-level) leadership within this stratified leadership structure.

[3] Data extracts from research project by the author with women doctoral advisors in arts, humanities and qualitative social sciences (2013–2018).

The thing pre-occupying me is my role as doctoral advisor for our department. This job has steadily grown, from handling inquiries, managing applications and organising provisional reviews at the end of the first year of candidature – those are still the three core tasks but each has got bigger and more difficult. I now have three PhD degrees to manage, rather than the one I began with. Most significantly, the Faculty restructuring three years ago – that abolished our department and made us part of a larger school – has disempowered us all and made all our administrative tasks more difficult. Rather than sending recommendations to decline or approve to my Head of Department, I now have to send them to my local disciplinary head, then the School Postgrad Director, so two steps instead of one. Also, I can no longer decline an application without first justifying it to the Associate Dean Postgraduate at Faculty level, who always pushes back and wants it rethought, or sent to other disciplines to consider. Part of this is driven by the University's desire to attract more international PhD students, which also means we are being sent applications from applicants with lower and lower GPAs. In terms of the first-year reviews, there is a new template form that we are/I am supposed to complete, which requires more work than in the past. Added to all of this are growing expectations that I will take a pastoral care role with students and run 'professionalisation' programmes for them – workshops on writing for publication etc. I really enjoy that side of the job in fact. It's the overall, ever-growing load and the difficulty of having our recommendations accepted that makes the job increasingly unattractive.

(Donna, Retreat 2017, Workshop 2)

Fig. 1 Doctoral advisor (Donna)

Their descriptions are worth including in full because they show the complexity of demands that impede creative leadership in doctoral education (Figs. 1 and 2):

From these accounts, we can see that mid-level leadership in doctoral education is often as compromised as the top-level, for similar reasons (personal reward, obedience, ambitious targets set to please those above) and that this directly affects what happens at the local level. We see the dark side of cascading institutional recruitment targets and what I have come to think of as the 'policification' of educational processes. Careful (and time-consuming) academic judgement is put aside in favour of meeting key performance indicators (KPIs) set above but enacted below. Creative academic leadership, including providing development opportunities for and with doctoral students, is displaced by laborious—and frequently mutating and proliferating—tasks of bureaucratic compliance, as Virginia laments when she wishes her role as PhD Director for her school could be other than form-filling.

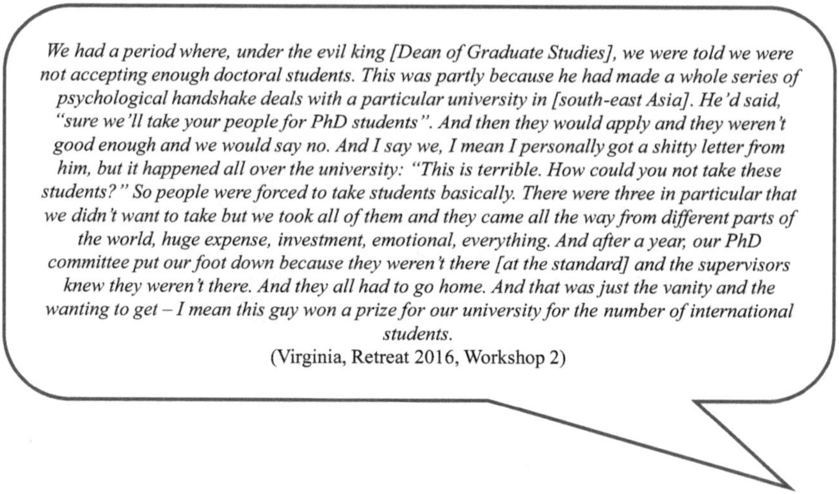

Fig. 2 Doctoral advisor (Virginia)

Fruitful Possibilities in Local Leadership

Yet the creative work of fostering research and scholarship is where the hearts of most academics remain. And so, for the rest of this chapter, I turn my attention to the 'lowly' (and largely overlooked) level of leadership that occurs within localised academic units (schools, departments and research centres) where students and supervisors interact together. The position of doctoral advisor (or director) for a school (or department or research centre) seems to me the most fertile ground for thinking about how leadership might foster researcher interdependence. My preference for paying attention to this level of leadership is informed by feminist theorisations of effective leadership (see, for example, Sinclair, 2014) as something different from the 'great man' model that imbues corporate university culture. These other modes of leadership pay attention to the ground of experience (of also being a doctoral supervisor, for example), to being in the midst of things that matter to us, that affect us and that cause concern and reflection. Where that concern and reflection become charged with a desire to intervene for change or, at least, to increase the possibilities for flourishing, we find prospects for meaningful leadership that entails "reflectiveness and *contestation*" (Sinclair, 2004, p. 8, my italics).

Thinking at this more hopeful, lively level of leadership, I want to address the idea of researcher interdependence. Like other authors in this book,[4] I find the idea of interdependence more practically and ethically compelling to advocate for with doctoral students (and colleagues indeed) than independence per se.[5] Interdependence holds together, if always in tension, ideas of independence (the I) *and* intersubjectivity (the we) as necessarily entangled for all human endeavours including academic research. While much is made of the need to cultivate independence in doctoral students as they move from the relative dependence of undergraduate and masters education, or as they move from one academic culture to another, I think this cultivation can take place within efforts to foster interdependence. It is a vision of growing with and through interactions with others, in which the student takes an active rather than passive role.

When Bill Green, in his 2018 essay *Addressing the curriculum problem in doctoral education*, cites Hannah Arendt to observe that an "extraordinary promise is arguably at issue in doctoral work—the promise of natality", he sees "futurity, becoming and the new" (p. 70). Something like this promise excites many who undertake doctoral study. But the promise that comes just as forcefully is that of intersubjectivity: there is no natality without parents, without the social. Whatever is new that emerges from doctoral education—the original thesis, the newly licensed researcher—is born from complex interplays with many others, living and dead. And so in the messy process of thesis and researcher being born, local leadership may generatively figure as midwifing that which is coming into being.

In my experience, the role of school doctoral advisor is a short-term service, often taken in rotation by academics[6] within an academic unit. It is usually considered a part of the normal service load that every academic is required to contribute and so not subject to extra income or other perks. The role may include some or all of the following: administrative responsibilities such as being abreast of the institution's regulations, access to information about all doctoral students within the unit and signing responsibilities at various key stages. It may also include pastoral

[4] See Chaps. 1, 3, 8, 9, 12, 13 and 15.

[5] Advocacy for interdependence may be more needed in relation to arts, humanities and social sciences research than in many arenas of STEM research, where it is an explicit condition of possibility.

[6] Also called (especially in North America) 'faculty'.

responsibilities for supporting students with personal and institutional issues that are hindering their progress, including supervision difficulties. And doctoral advisors may also have academic responsibilities. For example, in our school, the doctoral advisor has the roles of chairing all first-year doctoral student reviews[7] and, explicitly, of "fostering an active student research culture" (School doctoral advisor role description, 2016).

A commitment to encourage an active student research culture is the kind of umbrella under which fostering and strengthening doctoral students' understanding and experience of academic interdependence can shelter. Admittedly, many doctoral advisors' capacity for creative development work is sorely constrained in the modern university, as Virginia rues above, where KPIs and their tools (policies, numerical targets, ever-new online systems) and consequences (more and more make-work) constantly expand, where discretionary budgets are minimal or non-existent, where exhaustion and disillusionment with workplaces are rife. Moreover, doctoral students' interest in non-compulsory activities is fragile: it is easily eroded by part-time status, the need for paid work, the belief that there is nothing to learn from other students, feelings of pressure and exhaustion in relation to their own progress and so on. (See also Rainford's chapter stressing the value of interpersonal contact during the doctoral journey.) Most of these erosive forces have structural underpinnings: a pervasive culture of individualism, inadequate scholarships, strict (and arguably insufficient) time-to-completion norms and aggressive institutional mandates to put everything online, creating a culture where face-to-face is no longer the norm, thus reiterating over and over that being with others is certainly not essential for academic work and probably not important. And yet, as Chris Golde's (2005) research suggests, the culture of a doctoral student's academic unit matters, including *for* that student's persistence.

But where conditions allow, many quite concrete strategies—all of which are forms of hidden (informal) curriculum—can be used by the doctoral advisor to midwife student interdependence (this list is not exhaustive):

- Working with property services to cluster students physically together in convivial places with easy informal access to academics (shared tearooms, etc.).

[7] All doctoral students must pass through this step—presenting a written research proposal to an internal review panel—in order to progress in the doctoral programme.

- Normalising student co-representation on committees in the school, perhaps even having a graduate student committee, and partnering with reps to provide student-facing activities.
- Recruiting other colleagues to contribute to such activities. Students are often more interested when academics participate; another result can be a livelier department research culture.
- Encouraging students to rely on each other through participating in student-led writing groups, writing retreats and reading groups. Sometimes this might entail invited input from the doctoral advisor or other academics to assist with modelling what interdependence looks like in action. And procuring funds to support such initiatives because money announces their value.
- Teaching some of the skills of interdependence, for example, how to constructively review others' writing and give balanced feedback. (For example, see Makara et al.'s and Peseta et al.'s chapters.)
- Striking a thoughtful balance between providing what students want/believe is worthwhile and what the doctoral advisors themselves want/believe is worthwhile. (Students won't always know what they need to know and what will help them.)

- Promoting the virtue of interdependence explicitly: emphasising, when engaging with students, that this is what academic and other kinds of professional lives require to provide the best conditions for creative human flourishing. Acknowledge that interdependence is by no means a 'natural' attribute but that we can cultivate it. We can practise actions of goodwill and generosity, of dealing constructively with diverse and disagreeing points of view and of giving time to others and their work.

The role of the doctoral advisor here is to make the case to students that interdependence is not an optional extra to doctoral study but part of the long process of becoming someone who is able to work well with others, to appreciate what others have to offer, and to be able to offer ourselves. (Examples of creative ways in which interdependence is fostered can be observed in chapters by Dangeni et al., Guerin and Aitchison, Makara et al., Peseta et al., Rainford and Wisker.) Some students will resist, enjoying the culture of competitive individualism, or simply preferring to work in isolation, focusing only on their own goals, deadlines and lines of thought. While we need to be real about these and other limitations imposed by the institutional conditions of doctoral education, we can also be creative and hopeful within them.

Doctoral Leadership as Midwifing the New

The figure of the midwife may be culturally female, but lively—life-giving—leadership in doctoral education belongs to us all. The promise of natality offered to those who embark on this arduous journey refers to the new researcher (or scholar) subject being born *and* the new knowledge, both of which are intimately entangled and profoundly social. The midwife's work is to assist in this difficult passage. Midwifing doctoral study has many possible dimensions, but surely it is (at least) to inspire and encourage doctoral students by setting their eyes firmly on the prize to come, to bring them together to help and inspire each other through the long labour of the PhD, and to show them that many struggles are not theirs alone but are structurally embedded in the process of conceiving original academic knowledge. At its best, the leader-as-midwife is able to articulate a *vision* for what makes doctoral study, and academic work at large, worthwhile.

Declaration of Conflict of Interest No potential conflict of interest was reported by the author.

Acknowledgements My gratitude to three colleagues who gave permission for their comments to be included.

References

Golde, C. M. (2005). The role of the department and discipline in doctoral student attrition: Lessons from four departments. *The Journal of Higher Education, 76*(6), 669–700.

Grant, B. M. (2011). The bothersome business of curriculum in doctoral education. In E. Bitzer & N. Botha (Eds.), *Curriculum inquiry in South African higher education* (pp. 253–266). Sun Press.

Green, B. (2018). *Engaging curriculum: Bridging the curriculum theory and English education divide*. Routledge.

Haynes, J. (2009). Living PhD: Metaphors of research, writing and supervision. In T. Brown (Ed.), *The Doctorate: Stories of knowledge, power and becoming* (pp. 26–32). Escalate@Bristol.

Macfarlane, B. (2014). Challenging leaderism. *Higher Education Research & Development, 31*(1), 1–4.

Sinclair, A. (2004). Journey around leadership. *Discourse: Studies in the Cultural Politics of Education, 25*(1), 7–19.

Sinclair, A. (2014). A feminist case for leadership. In J. Damousi, K. Rubenstein, & M. Tomsic (Eds.), *Diversity in leadership: Australian women, past and present* (pp. 17–35). ANU Press.

Slaughter, S., & Leslie, L. L. (2001). Expanding and elaborating the concept of academic capitalism. *Organization, 8*(2), 154–161.

Nested Leadership in Research Education

Søren S. E. Bengtsen and Lynn McAlpine

Introduction

When Graduate School leadership, or research education leadership more broadly, is reported in the literature, it may focus on the structural implementation of policy concerning quality assurance or the professionalization of doctoral education (Andres et al., 2015; Elmgren et al., 2016). Such perspectives contribute to understanding how Graduate Schools, institutionally centralized units, have increased in structural size and complexity. However, this leadership focus can seem depersonalized, with its focus on the formation of structural hegemonies. So, perhaps not surprisingly, these forms of research education leadership have been linked to

S. S. E. Bengtsen (✉)
Aarhus University, Aarhus, Denmark
e-mail: ssbe@edu.au.dk

L. McAlpine
University of Oxford, Oxford, UK

McGill University, Montreal, QC, Canada
e-mail: lynn.mcalpine@mcgill.ca

D. L. Elliot et al. (eds.), *Developing Researcher Independence Through the Hidden Curriculum*,
https://doi.org/10.1007/978-3-031-42875-3_19

potential conflicts (Cassuto, 2015; Manathunga, 2005)[1] as they may not attend to departmental and individual practices.

An alternate, more inclusive view recognizes leadership as including supervisors and research education co-ordinators (Boud et al., 2014). This chapter focuses on these latter leadership roles, emphasizing variation in individual independent and interdependent engagement and achievement. This stance shows the formation of research education leadership within institutional structures as highly dynamic and varies greatly in relation to how individual research leaders understand and enact their institutional and academic agency (Ashwin et al., 2015), while integrating personal life and career intentions and values (Bengtsen & McAlpine, 2022). Individuals take different stances in relation to different policies, and these stances vary from narrator (setting out the institutional agenda) through enthusiast, critic, and receiver (rely on others and ask for guidance) (Ball et al., 2011)—or innovator, broker, deliverer/monitor, developer (of people), and integrator (selecting and integrating other roles through reflective practice; Boud et al., 2014).

In our view, the enactment of leadership is relational, interdependent, and contingent on individual choices and life aspirations, alongside networking, collaboration, and a collective awareness. Further, it varies from formal to informal, internally through externally focused, and desired spheres of influence. So, research leadership involves everyday departmental and collegial policy-practice entanglements—even for senior leaders creating institutional policy texts responsive to external regimes while approaching the more local lived experience of research education. We build such understandings on 'nested leadership', or 'lived leadership' (inspired from Aoki's (1993) notion of the 'lived curriculum'[2]), and an understanding of the hidden curriculum in doctoral education and the PhD as promoted by Elliot et al. (2020). Even though research education leadership does not, as such, have a curriculum (it is not an educational but a leadership practice), research education leadership still contributes to the doctoral curriculum through developing and sustaining the formal structures and informal cultures within which doctoral education becomes enacted in any institutional and national context. In this chapter, we are

[1] Many academics view such management aspects of leadership sceptically (Bolden et al., 2012).

[2] That is, not the curriculum-as-plan (formal structures), but the totality of the lived doctoral experience, including the 'extra-curricular', the tacit, and taken-for-granted.

striving to unpack some of these forms of tacit or personalized knowledge (Polanyi, 2015) in the practice of research education leadership.

Through two cases, we show how more local research education leadership may take different forms, ranging from engagement with formal structures and efforts to bend and change the institutional regimes from within the local structural reality (Beth) to informal community building to create a caring micro-cosmos in the margins and peripheries of the local structural reality (Anna). These two case examples emerged from a longitudinal study of the institutional entanglements of doctoral supervisors in mid- and late-academic careers (Bengtsen & McAlpine, 2022). The institutional context is a UK elite research-intensive university in which, over the last two decades, senior research education leaders formalized and implemented strategies for doctoral education and supervision practices in line with UK policy requirements—while endeavouring to be responsive to more local institutional practices in two ways: first, through consultations with senior faculty research education leaders to generate the policy text and, second, by giving subsidiary units, such as departments, freedom to interpret and translate these policies into local practices. Our argument shows that research education leadership is not personal or institutional independent (or isolated) but an interdependent (or nested) practice. Research education leadership practices strive both to interconnect curricularly the two 'legs' of the PhD, research *and* education, and to interconnect institutionally across the various organizational layers of Graduate School leadership, PhD programmes, faculties, departments, and individual doctoral supervisors and scholars.

Theoretical Anchoring

The argument embedded in the two case-examples rests on understandings of 'nested leadership' drawn from McAlpine and colleagues' work on doctoral education taking place within a series of nested contexts (McAlpine & Amundsen, 2018; McAlpine & Norton, 2006) in which individuals' experiences, thinking, and action interact with structural constraints and affordances. Agency for individual research education leader is enacted within the dynamic interrelations of the opportunity structures of the system and the horizons for action of the individual—their preferred spheres of influence. In this view, leadership of research education, as with any other role, always takes place within the individual life-world in which work is embedded in nested contexts (local disciplinary communities, the surrounding departmental and institutional structures, and national/global structures and trends). Agency represents the extent to which individuals articulate and progress towards personal and work intentions and goals while navigating supporting and constraining structures; this means choosing how they participate, including modifying or refusing to participate in institutional structures (McAlpine & Amundsen, 2018). Thus, leadership in research education cannot be reduced to either structural determinism or arbitrary personal idiosyncrasy, but is constituted by the interdependence and interplay of individual decision-making, disciplinary, and organizational culture, and the formal structural policy-making of the institution. In this view, research-education leadership practices interconnect the various institutional sub-units and sub-cultures in research and educational environments across PhD programmes, departments, and smaller research teams (and even supervisory dyads) that constitute the lived curriculum of doctoral scholars and others involved in doctoral education. Research education leadership is, thus, a form of 'institutional knitting' together of various institutional contexts, often hidden or less visible within each context experienced separately.

Further, the argument draws from conceptualizations of academic and Graduate School leadership as the need to "merge formal and informal educational and learning spaces", to "link individual researcher trajectories with communal research environments", and to "connect institutional and societal domains and life-worlds" (Bengtsen, 2021, pp. 25–26). As also argued by Elliot et al. (2020), the institutional ethos and legitimacy of academic leadership hinges on community presence and engagement. In order for more senior and formally assigned leaders of research education to be perceived as actual leaders requires their initiatives and actions be recognized and acknowledged by the upper leadership forums through more local designated leaders to supervisors and doctoral scholars. As defined in a recent conceptualization of academic leadership (Bengtsen, 2022, p. 140), the interdependent relational quality of any formal institutional leadership agency becomes visible "where leadership becomes less associated with policy and strategy" and more strongly associated with an "ethical awareness both within and outside institutions and community building practices". Such nested leadership happens when leaders in researcher education infuse meaning, purpose, and educational vision into the opportunity structures of PhD programmes and Graduate School systems and attempt to address the constraints. As ongoing research shows (McAlpine et al., 2022), the purpose and vision, perhaps somewhat surprisingly, are not linked only to the strategic and institutional policy but are also channelled from personal PhD and research experience, informal collegial mentoring, and networking with external partners. Nested leadership is personal, life-world-based, and organizationally entangled—it is an interdependent form of institutional leadership practice.

Structural Change from Within: The Case of Beth

Beth is a professor in the field of language within the humanities faculty. She is of British descent, in her late 50s, and has been employed at her current university for over two decades. From early on in her career at her present university, Beth engaged actively with the formal institutional structures and endeavoured to bend and reshape them according to how she saw the necessity for new initiatives, for instance, around workshops and courses for doctoral scholars, co-supervision opportunities, and well-being and mental health issues in doctoral education. She generally views the institutional doctoral education policies as positive—though she is experiencing a tension between the expected and required individual autonomy in doctoral work and the increasing managerial approach to the PhD.

Recently, Beth has taken up the role of Head of Department, a close-to-full-time leadership position, so she will not teach for the duration of the appointment. She is also leading a strategy review. Beth, a highly experienced doctoral supervisor, has experience with different forms of co-supervision and collaborative doctoral work, including supervisors from external contexts. Engaging in various forms of supervision has led Beth to a less prescriptive and "dominating" approach to supervision, which she herself ascribes to her greater experience with co-supervision with colleagues in her university and external partners. Beth has felt for some time that a more collegial and collectively informed understanding of supervision practices is needed between supervisors and doctoral scholars in her department. During her work as a doctoral supervisor, before moving into her present formal leadership position, Beth began to take initiatives in her department and to co-develop programmes for doctoral scholars focusing on generic skills and knowledge about being a doctoral scholar at her university. Over time, Beth has also experienced some more difficult sides of doctoral supervision: getting doctoral scholars to take ownership and responsibility for their own PhD and doctoral scholars' well-being and mental health issues. Her focus on struggling doctoral scholars is carried over in her recent work on a strategic review committee regarding attainment gaps in race and culture.

Beth has been able to work and align with the present structural and bureaucratic hierarchies and decision-making processes, and she has put herself forward, if required, to take on formal leadership tasks herself in order to change the system and structure from within. Still, these formal leadership roles have taken up much time, thus decreasing her time to supervise and conduct research herself. She has become aware that formal leadership positions will not leave much opportunity for book writing and pursuing research agendas. So she will need to find balances not only for the doctoral scholars (and supervisors) she is developing new formal structures for but also for herself if she wishes to supervise, teach, and conduct research while being a formal institutional leader.

What we see in Beth's case is someone who takes on more formal leadership roles within the sphere of influence of her department, a context in which she has over time also shown informal research education leadership. Her work is relational, involving everyday collegial practices and entanglements in the service of **all** doctoral scholars (and supervisors) in the department, set within broader institutional policy goals. She recognizes that this commitment influences the other kinds of work she might like to do. Thus, we consider her an integrator (Boud et al., 2014), an individual who seeks to incorporate aspects of the range of policy actor roles through reflective practice.

Informal Community Building: The Case of Anna

Anna is a professor in the field of aesthetics, also within the humanities faculty. She is of British descent, in her late 40s, and has been employed at her current university for over a decade. Over the years, Anna has had very good success with building strong informal communities for her doctoral scholars and postdocs. She has held regular reading and discussion group meetings, which gather around six or eight of the current doctoral scholars and postdocs working with her. In the meetings, the group discusses a given paper or book key to the field, one chosen in turn by the group members individually. Anna takes on a facilitating role but otherwise strives for equality and democracy in the group, even though the members differ greatly in experience and formal institutional status. For Anna, it has been important to give everyone in the group a voice of equal weight and value. After the meetings, the group often goes to the pub and sometimes continues their discussions more informally over a drink or a meal, and talks about more private non-academic matters. The group has not existed in formal structural reality, so doctoral scholars or postdocs outside the group can only join if they learn about it informally and seek to join. Anna mentions that she has kept it that way intentionally so that she could create her own social and academic sustainability in the group by putting people together she felt would complement each other well intellectually and emotionally.

For both professional and personal reasons, Anna has recently been on a longer sabbatical, and upon her return, she realized that some of her efforts to encourage doctoral scholars to work more in teams seem to have disappeared; further, the ideas about collectivity in doctoral work within the department before she left have not been taken on board. Anna has also felt a growing isolation within her local disciplinary community due to her topic being rather marginal to the department's main priorities. She feels as well that, while she was away, doctoral education became more career-oriented, more competitive, less free, and creative. Given Anna's preference to prepare individuals for academic careers, her response has been to continue her own informal and structurally invisible supervision and mentoring: putting even more focus on the importance of social and emotional support in the group in order to balance the managerial approach. She can do this because these individuals wish to engage in her initiative. As noted above, Anna had tried through persuasion (informal leadership) to influence departmental practice but has avoided pursuing formal leadership roles and responsibilities when they became available. She works to resist the institutional bureaucracy in the ways she can to maintain as much humaneness as possible in a system she views to leave less and less room for doctoral scholars to include their existential and emotional lives in the PhD study process.

Anna's research education leadership is informal and institutionally less visible (partly due to Anna's own efforts to keep it that way), as would be the case with many supervisors. Her leadership vision rests more on an ethical awareness than a will to change the structural reality of the institution: choosing to avoid formal leadership roles given the policies are inconsistent with her own values. Given her leadership role is with her own doctoral scholars and postdocs, it remains relational, but (in contrast with Beth) the sphere of influence is directed **specifically** at those she knows and wants to support. Her stance towards the policies is that of a critic (Ball et al., 2011), but she still values informal leadership and is willing to take it on where it does not compromise her values.

Nested Leadership in Doctoral Education

The cases illustrate that leadership in research education is a dynamic process of change-making influenced by the career and life-trajectory of the individual academic assuming formal or informal leadership in doctoral education. Thus, leadership in research education is much broader and more complex than often envisaged, manifesting itself in multiple ways as individuals negotiate personal goals and visions for academic culture, often with a rootedness in personal values and life experiences. As McAlpine and Åkerlind (2010) have argued, the cases illustrate a process of becoming an academic, which is not a stable and fixed role and practice but a flexible, plastic, evolving one. Also, as argued by Barnett and Bengtsen (2020), knowledge and leadership practices may be viewed as life-forms and infused with life within the academic structures delimiting them. Leadership in research education is a nested practice that takes place in the intersections of personal, academic, and structural contexts and realities. Leadership agency is at the same time dependent on the evolving structural opportunities *and* the resistance, interpretation, dodging, or changing of such structures. As seen in the case examples, institutional structures can be bent and reshaped through initiative and engagement (Beth), or they may be bypassed in order to build and sustain academic communities in the in-between-spaces of institutional reality (Anna). The interdependent, or nested, nature of the leadership practices shows the constant institutional dialogues (which sometimes break down or are rekindled and renegotiated) happening with and across various institutional layers *through* the lived leadership of Anna and Beth.

The cases also illustrate that leadership in research education does not rest with Heads of Graduate Schools alone, or Directors of PhD programmes, but is enacted ongoingly through career trajectories from early to mid- and late-career. Also, leadership does not merely hinge on the formally assigned leaders but is being practised by the ones who demonstrate leadership agency through their supervision practices and cohort mentoring. Leadership is highly person-dependent in that it requires someone who has a vision, a goal, and a will to change something—hopefully for what they think is better. (Such types of leadership have been featured in Makara et al.'s Journal Club, Preece's multiple ways of doing and Wisker's Fridaying chapters.) Agency often depends as well on the will to go through periods of tiredness but with the resilience to return to the leadership tasks anew. Finally, leadership is highly interdependent in the sense that it never rests solely on the individual but on the community and sustainability of informal milieus, as in Anna's case, as well as, as in Beth's case, institutional and collegial recognition in order to change the structural lived reality of the institution. The interdependence shows there are many roads to leadership—further, that it is possible and worthwhile to continue striving for even better institutional and doctoral education futures.

Declarations of Conflicts of Interest Nothing to declare.

Acknowledgements The writing of the chapter was made possible through the project 'Research for impact—integrating research and societal impact in the humanities PhD', which is a Sapere Aude research project funded by the Independent Research Fund Denmark (DFF). Also, the earlier data collection on which the chapter is based was made possible through a travel grant from the Aarhus University Research Foundation (AUFF).

References

Andres, L., Bengtsen, S., Crossouard, B., Gallego, L., Keefer, J., & Pyhältö, K. (2015). Drivers and interpretations of doctoral education today: National comparisons'. *Frontline Learning Research, 3*(2), 63–80.

Aoki, T. (1993). Legitimating lived curriculum. Towards a curricular landscape of multiplicity, poetic, phenomenological and hermeneutic discourse in which life is embodied in the very stories and languages people speak and live. *Journal of Curriculum and Supervision, 8*(3), 255–268.

Ashwin, P., Deem, R., & McAlpine, L. (2015). Newer researchers in higher education: Policy actors or policy subjects? *Studies in Higher Education, 41*(12), 2184–2197.

Ball, S., Maguire, M., Braun, A., & Hoskins, K. (2011). Policy actors: Doing policy work in schools. *Discourse: Studies in the Cultural Politics of Education, 32*(4), 625–639.

Barnett, R., & Bengtsen, S. (2020). *Knowledge and the University. Re-claiming Life*. Routledge.

Bengtsen, S. (2021). Agency in doctoral education. Towards Graduate school cohesion and a heightened societal awareness. In A. Lee & R. Bongaardt (Eds.), *The future of doctoral research. Challenges and opportunities* (pp. 18–27). Routledge.

Bengtsen, S. (2022). Cultural leadership. In R. In Barnett, S. Bengtsen, & R. T. Nørgård (Eds.), *Culture and the university. Education, ecology, design* (pp. 137–148). Bloomsbury.

Bengtsen, S., & McAlpine, L. (2022). A novel perspective on doctoral supervision. Interaction of time, academic work, institutional policies, and lifecourse. *Learning and Teaching in the Social Sciences (LATISS), 15*(1), 21–45.

Bolden, R., Gosling, J., O'Brien, A., Peters, K., Ryan, M., & Haslam, A. (2012). *Academic leadership: Changing conceptions, identities and experiences in UK higher education.* Leadership Foundation for Higher Education.

Boud, D., Brew, A., Dowling, R., Kiley, M., McKenzie, J., Malfroy, J., Ryland, K., & Solomon, N. (2014). The coordination role in research education: Emerging understandings and dilemmas for leadership. *Journal of Higher Education Policy and Management, 36*(4), 440–454.

Cassuto, L. (2015). *The graduate school mess. What caused it and how can we fix it.* Harvard University Press.

Elliot, D., Bengtsen, S., Guccione, K., & Kobayashi, S. (2020). *The hidden curriculum in doctoral education.* Palgrave Macmillan.

Elmgren, M., Forsberg, E., Lindberg-Sand, Å., & Sonesson, A. (2016). *The formation of doctoral education.* Lund University.

Manathunga, C. (2005). The development of research supervision: "Turning the light on a private space". *International Journal for Academic Development, 10*(1), 17–30.

McAlpine, L., & Åkerlind, G. (2010). *Becoming an academic. International perspectives.* Palgrave Macmillan.

McAlpine, L., & Amundsen, C. (2018). *Identity-trajectories of early career researcher: Unpacking the post-PhD experience.* Palgrave Macmillan.

McAlpine, L., & Norton, J. (2006). Reframing our approach to doctoral programs: A learning perspective. *Higher Education Research and Development, 25*(1), 3–17.

McAlpine, L., Bengtsen, S., & Gibson, A. (2022—submitted). *Graduate school leadership: How Danish humanities research education leaders constitute their roles.* Submitted symposium for the 20th Biennial EARLI Conference 'Education as a hope in uncertain times', Aristotle University of Thessaloniki, Greece.

Polanyi, M. (2015). *Personal knowledge. Towards a post-critical philosophy.* University of Chicago Press.

PART V

Insights on Doctoral Education Beyond Academia

Putting the 'Extra' in Extracurricular: Why Going Off-Script is Important for Life After the Doctorate

Billy Bryan

The Case for Extracurricular Learning in Doctoral Education

Involving doctoral scholars in 'extracurricular' activities outside of their direct area of study may be interpreted as 'going off-script'. Additional activity might seem like an unnecessary addition to their to-do lists. But the 'extra' does not have to mean 'burdensome' or 'surplus to requirement'. There are hidden benefits to extracurricular activities that can enrich the doctoral experience, build researcher independence and help graduates secure the jobs they want.

Extracurricular development opportunities are, by definition, part of the 'hidden curriculum' in that they are neither compulsory for degree completion nor discoverable and accessible to all doctoral scholars. They

B. Bryan (✉)
RAND Europe, Cambridge, UK
e-mail: bbryan@randeurope.org

D. L. Elliot et al. (eds.), *Developing Researcher Independence Through the Hidden Curriculum*,
https://doi.org/10.1007/978-3-031-42875-3_20

are opportunities that can be used to extend pedagogical practices (e.g. writing retreats that embed academic and professional communication skills) and support personal, social and professional development needs (e.g. volunteering as part of a school's outreach initiative). Extracurricular learning is ultimately driven by the doctoral scholar but facilitated by 'hidden curriculum agents', such as employers and doctoral developers. A differentiating factor of extracurricular learning is that the diversity and range of those agents can, and arguably should, be sizable and from both academic and non-academic sources.

Demonstrating independence is more than the ability to manage a doctoral research project. In my view, it is the ability to critically appraise the world around us and to develop intelligent questions and robust methods of answering them. Our capacity to self-regulate (actively driving our thoughts and actions towards our goals) underpins this (Wille, 2020), comprising elements such as strategic planning, self-monitoring and reflexivity (Bryan, 2017). (This is aligned with Elliot's chapter on developing a metacognitive disposition.) These critical skills are increasingly seen as valuable by employers outside of academia who employ most doctoral graduates immediately after their studies (Hancock, 2021). The fact that a range of academic and non-academic skills is necessary for researcher development has been recognised at the international national level since at least 2005. Doctoral development frameworks and principles in the UK and Europe emphasise preparing graduates for non-academic careers, including domains such as enterprise, policy and career management, in addition to core research skills (Christensen, 2005; Vitae, 2011). Those are the same skills described as lacking in doctoral programmes in Australia, the UK and Europe (Nerad, 2015) and in doctoral graduates by non-academic employers (Wille, 2020).

At academic department level, the typical doctoral programme still tends to assume that every doctoral scholar wishes to become an academic, neglecting to adequately prepare them for a range of career paths, or positioning such careers as 'alternative' implying they are for those who fail to succeed in academia (Sharmini & Spronken-Smith, 2019). This leaves the doctoral scholar (and doctoral development staff) to parse the hidden curriculum to work out what careers are out there, what skills and experience they require and how to access them.

This chapter provides doctoral developers, supervisors and doctoral scholars with a guide to what extracurricular development opportunities mean in the doctoral context and how doctoral scholars can be empowered to reap the benefits. There is comparatively little research or practical guides on the benefits of extracurricular learning and how doctoral scholars can access them, compared to the undergraduate context in which employability and experiential benefits are well understood (Stuart et al., 2011; Thompson et al., 2013). This chapter draws upon the notable exceptions in the literature, including student-led case studies, such as Sum (2022).

What Are 'Extracurricular Development Opportunities,' and What Is There to Choose From?

Put simply, extracurricular development opportunities are any personal and/or professional development activities outside the core doctoral development curriculum and are not required to meet the criteria for achieving a doctoral degree. Extracurricular development opportunities are a way of both learning about and better preparing doctoral scholars for a range of non-academic careers. Extracurricular learning is not just about playing on a sports team or doing extra work on a supervisor's side project. In fact, any activity done alongside the doctoral project can be 'extracurricular'. Important skills are built for life and work post-graduation, particularly improved self-regulation (i.e. independence) by planning, managing and reflecting on activities outside the doctoral project, which non-academic employers recognise as valuable (Wille, 2020).

The paragraph above alludes to how extracurricular learning can contribute to and even help exceed the core doctoral curriculum, for example, teamwork and influencing via academic committees and argument construction through a speakers club. Their relevance and utility to the doctoral experience are largely clear. The degree to which those activities are 'hidden' is what the remainder of this chapter aims to deconstruct by making clear what extracurricular development opportunities are and how they can be made to work in favour of doctoral scholars.

I have conceptualised three (non-exhaustive and interrelated) groups to describe the kinds of extracurricular learning a researcher might do:

- **Doctorate adjacent**: There are a suite of activities that doctoral scholars participate in that are not strictly part of their studies but have become synonymous with traditional doctoral education approaches. These are by far the most common and accepted (to developers, supervisors and doctoral scholars) extracurricular development opportunities that are *not* required to complete a doctorate, apart from publishing articles, which is mandatory in some countries/programmes (e.g. in Latvia).

  - Examples: publishing knowledge products (e.g. journal articles, monographs, creative outputs), reviewing journal submissions, sitting on departmental committees, organising and attending conferences, organising journal and writing groups and exhibitions—there are many existing online guides for this type of activity (Minnis, 2019).

- **Personal and professional side-hustles:** These extracurricular development opportunities are most useful for broadening career horizons, meeting people outside the doctoral community and filling skills gaps. Some of these examples are not optional/extracurricular for all doctoral scholars (e.g. caring for a dependent, paid work to cover study costs) and it is worth recognising here the hard work and skill required to consistently perform those responsibilities alongside doctoral study.
  - Examples: part- or full-time work/care responsibilities and domestic labour, paid or unpaid internships outside academia, secondments, industry placements, research assistant roles, public engagement activities, school outreach, teaching, supplemental qualifications and professional recognition, writing for a magazine or blog.
- **Social and community:** The benefits here include developing a sense of community, a suite of skills (global citizenship, equality and diversity) and meeting yet more people outside a doctoral scholar's usual circles. These offer multiple and varied benefits. For example, student groups are particularly helpful not only for academic purposes (e.g. conference organising) but are arguably most useful for building communities and fostering digital well-being in what can be a lonely doctoral scholar journey (for a practical example of a 'PhD Society', see Sum, 2022).
  - Examples: volunteering at a food bank, campaigning at the students' union, fundraising for charity, organising in the community and internationally, taking part in clubs and hobbies, doctoral scholar student-led advocacy groups and speakers clubs.

Empowering Doctoral Scholars to Reap the Benefits of Extracurricular Learning

Postgraduate researchers can hugely benefit from 'hidden curriculum' experiences which support, empower and enable them to succeed during and after their doctorates. Research into how graduates derive value from their doctorates found that extracurricular experiences helped prepare doctoral graduates to secure non-academic jobs and build skills they might not otherwise have gained in their doctoral programme (Bryan & Guccione, 2018). A follow-up study confirmed with a larger population that personal and career doctoral value domains had the most impact on overall value judgements, meaning that extracurricular learning can potentially underpin to what extent people value their doctorates, or not (Guccione & Bryan, 2023). Participation in extracurricular activities can also bring wider academic, social and wellbeing benefits, and help doctoral scholars feel like they 'belong' in their academic community (Winstone et al., 2020). Extracurricular learning can help combat loneliness and mental health difficulties, which are increasingly prevalent in doctoral populations (Guthrie et al., 2017).

Most extracurricular development opportunities allow the doctoral scholars to meet and work with new people that they otherwise may not have engaged with. Incidental and infrequent interactions, such as informal conversations with peers, friends and role models, can be just as significant as formal and long-term interactions (e.g. with a manager during an industry placement). Such relationships are strong enablers in the transition to research independence (Baker & Pifer, 2011), which is itself enabled by being part of the wider doctoral community (Gardner, 2008), and can pay dividends when securing roles post-graduation (Bryan & Guccione, 2018).

However, access to extracurricular activities is not equal (Winstone et al., 2020). Researchers who study part-time, remotely, have caring responsibilities, are disabled or chronically ill, or who self-fund their studies are less likely to have the time and resources to participate in any activity beyond their core studies. Immigrants, women, people of colour and people with disabilities, do not enjoy the same level of access and comfort in academic and professional spaces, or have the same views on their own research independence (Blaney et al., 2022), as what is thought of as the 'typical' doctoral candidate: young, white and in receipt of public or private scholarships.

Beyond access, each doctoral scholar will have varying learning needs and motivations depending on where they are in their programme, particularly 'mature' learners and those in non-traditional doctoral programmes (e.g. EdD, MD) (Radda, 2011). For example, a part-time doctoral scholar in their fifth year, working full-time in their desired role outside academia and far away from campus, may not be interested in attending employability seminars but may be grateful for a fortnightly evening writing retreat to finish their thesis chapters and to access peer support. (For a discussion on part-time doctoral scholars' needs, see Rainford's chapter.) A doctoral developer should consider the whole spectrum of their cohort when looking to empower doctoral scholars to participate in the extracurricular development opportunities that fit their needs.

The rich variety of extracurricular development opportunities is both a strength and a risk. Not all opportunities are made equal, and the fact that they tend to exist outside of the institutional remit means that the doctoral developer may have less knowledge of their utility and relevance. The developer's role is to empower doctoral scholars to make independent decisions about participation and to create and facilitate opportunities that are possible under the institution's remit (e.g. funding for travel grants, volunteer days, cross-departmental networking). This harkens back to improving doctoral scholars' self-regulation and independence with regard to personal and professional development.

In formulating a programme of extracurricular support, doctoral developers and supervisors should include the following design features:

Staff-Doctoral Scholar Partnerships

Staff and doctoral scholars planning and implementing extracurricular learning together is not only an effective model for engagement (Nachatar Singh, 2018), but it also reinforces the idea to everyone involved that academic staff are hidden curriculum agents as well as being responsible for what is considered the formal doctoral development curriculum. Supervisors (and other direct academic colleagues) are the most influential actors in the doctoral journey. Their building in time for, or helping to organise, extracurricular activities gives implicit permission to researchers to engage in professional and personal development. This helps to position extracurricular learning as a genuine site of pedagogical practice for doctoral education, rather than an additional or unnecessary element. It also helps doctoral scholars become more independent in advocating for their own learning needs, accessing support where they need it and having joint-accountability for their development.

Benefits can be multiplied and democratised in the case of student-led extracurricular learning. PhD advocacy groups (e.g. doctoral scholar student societies under students' unions) help facilitate access to extracurricular learning, create doctoral scholar communities (reducing loneliness, particularly for remote students, meeting social and psychological needs), and give members the opportunity to practice their governance, organisational and policy skills (Sum, 2022)—all success factors for researcher independence. Staff support and encouragement for these groups should also be provided as part of the whole suite of doctoral development interventions.

Strategies include doctoral scholar representation on academic and community forums; industry placements that are actively supported by supervisors; co-led doctoral scholar-staff training sessions, doctoral scholar representative/ambassador programmes (NB: these activities should not exploit the labour of doctoral scholars and remuneration should be offered); mutually agreed and co-created development plans that feature extracurricular learning with regular monitoring and support; facilitation of doctoral scholar societies and committees.

Be Driven by the Needs of Postgraduate Researchers

Doctoral scholars should be empowered to lead and/or inform the kinds of extracurricular development opportunities offered to them. This will help empower doctoral scholars to create pathways and harness a tailored hidden curriculum based on personal needs and professional aspirations. Simple surveys combined with ongoing structured or informal dialogue should take place to explore what doctoral scholars are interested in, what skills they want to develop and what would motivate them to participate. Effective dialogue also helps identify access issues. 'Discovery'-type sessions (e.g. interactive panel discussions with doctoral graduates in different sectors) should also be offered for those who want to develop but do not know in what area or how they might do it.

Researchers being enabled to lead or co-lead the activities helps to build leadership and organisational skills, and is likely to make the activity more relevant to their needs as well as their resumes. More importantly, this process will help increase their independence and self-regulation skills, such as task planning and self-evaluation.

Strategies include doctoral scholar-led departmental fora and consultation; routine departmental review of training needs analysis or annual planning documents; alumni careers seminars showcasing sectors/occupations of interest; regular departmental forums on current topics; mentoring; doctoral scholar exit surveys or interviews; familiarisation with the latest literature and best practice on extracurricular development opportunities in doctoral education.

Have a Diverse Menu of Accessible Opportunities

Development needs will naturally be different for each individual student, and the extracurricular offering should reflect that, offering support for academic, personal, social and psychological needs. Career, academic, leadership skills and social and network-building activities should feature equally, as not all doctoral scholars are completely career-minded nor do they all want to attend social gatherings. Equally, not everyone can attend events in the evening or at lunchtime, or feel comfortable in purely academic spaces dominated by groups they cannot relate to. Regardless, opportunities for extracurricular learning should be communicated equitably (Blaney et al., 2022), with extra care taken to reach and encourage those whose backgrounds and circumstances might mean they are less likely to participate.

Strategies include rotating the days and times of activities; creating specific self-paced online learning resources; circulating videos and notes from each activity; using different accessible venues and different online platforms; highlighting a range of different experiences of study, career choice and employment; ensuring widely advertised opportunities with extra targeting for underrepresented doctoral scholar groups.

Staying 'Off-Script'

Throughout the different highs and lows of the doctoral journey, most doctoral scholars will be looking to meet like-minded people, try something new, connect with new networks and improve their chances of securing their desired role after graduation. They will learn to better self-regulate if they are empowered to participate in extracurricular learning: identify what their development needs are and how to address them now and post-graduation. Their ability to independently explore 'off script' opportunities will serve them well in their lives post-graduation. (Securing jobs doctoral scholars want post-PhD was also highlighted in Aarnikoivu's, Dai's and Dangeni's chapters.)

Extracurricular opportunities and the support to access them should be permanent and inclusive features of all doctoral development programmes. It is worth repeating that inequalities inherent in doctoral education (e.g. having university educated parents) can mean that some students access extracurricular benefits more readily than others, leading to those who miss out being less aware of, and well prepared for, post-graduate careers (Blaney et al., 2022). Support from supervisors in balancing core academic work and extracurricular learning is crucial here to avoid regret over opportunities missed and to guard against burnout (Winstone et al., 2020). We should also be wary of promoting 'tyranny of engagement', which imposes a restrictive, westernised notion of what constitutes 'acceptable' participation, rather than an empowerment model of engagement (Gourlay, 2015).

Doctoral programmes can better equip doctoral scholars to take up non-academic careers as those programmes play a key role in setting overall career expectations, including clarifying the value of the doctorate beyond the academy (Aarnikoivu et al., 2019). This makes a difference in the long run, as the perceived value of the doctorate has been strongly linked to the fulfilment (or not) of career expectations post-graduation (Guccione & Bryan, 2023). Taking this a step further, there must be a stronger feedback loop between those facilitating entry to doctoral studies, those developing doctoral scholars and those employing them (Bryan, 2023).

It is time to take advantage of the hidden curriculum of doctoral development by making extracurricular learning the norm for researchers and all those involved in guiding them towards life post-doctorate. This transition to giving doctoral scholars the agency and structure to plot out their own development journey will contribute to their own research independence, the sense of community and belonging in the university community at large and send graduates out into the world with open minds to tackle intractable challenges.

Declarations of Conflicts of Interest ***The authors declare the following competing interests: Dr Billy Bryan is currently serving as an Evaluation and Research Leader at RAND Europe; however, the views, opinions, findings, conclusions and recommendations contained herein are the author's alone and not those of RAND Europe or its research sponsors,***

clients, or grantors. He is also a trustee at the University of Sussex Students' Union, a member of the UK Government's Evaluation and Trial Advice Panel (ETAP), and an advisor at the Healthy Heart Trust.

Acknowledgements n/a

References

Aarnikoivu, M., Nokkala, T., Siekkinen, T., Kuoppala, K., & Pekkola, E. (2019, January 11). Working outside academia? Perceptions of early-career, fixed-term researchers on changing careers. *European Journal of Higher Education, 9*(2), 172–189. https://doi.org/10.1080/21568235.2018.1548941

Baker, V. L., & Pifer, M. J. (2011, March). The role of relationships in the transition from doctoral student to independent scholar. *Studies in Continuing Education, 33*(1), 5–17. https://doi.org/10.1080/0158037x.2010.515569

Blaney, J. M., Wofford, A. M., Jeong, S., Kang, J., & Feldon, D. F. (2022, June 15). Autonomy and privilege in doctoral education: An analysis of STEM students' academic and professional trajectories. *The Journal of Higher Education*, 1–27. https://doi.org/10.1080/00221546.2022.2082761

Bryan, B. (2017). *The impact of two different feedback models on the immediate and future learning strategies of medical undergraduates* [Doctoral thesis, University of Sheffield] White Rose eTheses Online. https://etheses.whiterose.ac.uk/18933/

Bryan, B. (2023, April 6). It is time to revamp our approach to doctoral careers. *Nature Human Behaviour*. https://doi.org/10.1038/s41562-023-01590-0

Bryan, B., & Guccione, K. (2018, May 30). Was it worth it? A qualitative exploration into graduate perceptions of doctoral value. *Higher Education Research & Development, 37*(6), 1124–1140. https://doi.org/10.1080/07294360.2018.1479378

Christensen, K. K. (2005). Bologna Seminar Doctoral Programmes for the European Knowledge Society. *General Rapporteur's Report. Salzburg*, 3–5. Retrieved October 15, 2022, from https://eua.eu/component/attachments/attachments.html?task=attachment&id=1881

Gardner, S. K. (2008, May). "What's too much and what's too little?": The process of becoming an independent researcher in doctoral education. *The Journal of Higher Education, 79*(3), 326–350. https://doi.org/10.1080/00221546.2008.11772101

Gourlay, L. (2015, March 13). 'Student engagement' and the tyranny of participation. *Teaching in Higher Education, 20*(4), 402–411. https://doi.org/10.1080/13562517.2015.1020784

Guccione, K., & Bryan, B. (2023, January 2). Worth doing but not worth having? The influence of personal aspirations and career expectations on the value of a doctorate. *Studies in Graduate and Postdoctoral Education, 14*(1), 83–98. https://doi.org/10.1108/SGPE-02-2022-0012

Hancock, S. (2021, January 4). What is known about doctoral employment? Reflections from a UK study and directions for future research. *Journal of Higher Education Policy and Management, 43*(5), 520–536. https://doi.org/10.1080/1360080x.2020.1870027

Minnis, S. E. (2019, June). Doctoral juggling: Making time for scholarship. *New Horizons in Adult Education and Human Resource Development, 31*(3), 78–81. https://doi.org/10.1002/nha3.20259

Nachatar Singh, J. K. (2018, February 13). Evidence and benefits of postgraduate international students-staff members partnership in extra-curricular activities: A Malaysian perspective. *Higher Education Research & Development, 38*(7), 1475–1488. https://doi.org/10.1080/07294360.2018.1436527

Nerad, M. (2015, March 1). Professional development for doctoral students: What is it? Why now? Who does it? *Nagoya Journal of Higher Education, 15*, 285–319. Retrieved October 15, 2022, from https://www.cshe.nagoya-u.ac.jp/publications/journal/no15/14.pdf

Radda, H. (2011, November 30). ERIC—EJ980170—From theory to practice to experience: Building scholarly learning communities in nontraditional doctoral programs. *InSight: A Journal of Scholarly Teaching, 2012.* Retrieved October 16, 2022, from https://eric.ed.gov/?id=EJ980170

Sharmini, S., & Spronken-Smith, R. (2019, December 13). The PhD—Is it out of alignment? *Higher Education Research & Development, 39*(4), 821–833. https://doi.org/10.1080/07294360.2019.1693514

Stuart, M., Lido, C., Morgan, J., Solomon, L., & May, S. (2011, October 19). The impact of engagement with extracurricular activities on the student experience and graduate outcomes for widening participation populations. *Active Learning in Higher Education, 12*(3), 203–215. https://doi.org/10.1177/1469787411415081

Sum, K. (2022, July 11). Supporting doctoral researchers' social isolation through a student-led community *The Journal of Educational Innovation, Partnership and Change*. Retrieved October 16, 2022, from https://journals.studentengagement.org.uk/index.php/studentchangeagents/article/view/1098

Thompson, L. J., Clark, G., Walker, M., & Whyatt, J. D. (2013, April 10). 'It's just like an extra string to your bow': Exploring higher education students' perceptions and experiences of extracurricular activity and employability. *Active Learning in Higher Education, 14*(2), 135–147. https://doi.org/10.1177/1469787413481129

Vitae Researcher Development Framework (RDF). (2011). Vitae website. Retrieved October 16, 2022, from https://www.vitae.ac.uk/vitae-publications/rdf-related/researcher-development-framework-rdf-vitae.pdf/view

Wille, L. (2020). *PhD holders through the eyes of non-academic employers: A state-of-the-art literature review*. Department of Work, Organisation and Society. Retrieved October 15, 2022, from https://biblio.ugent.be/publication/8683644

Winstone, N., Balloo, K., Gravett, K., Jacobs, D., & Keen, H. (2020, February 29). Who stands to benefit? Wellbeing, belonging and challenges to equity in engagement in extra-curricular activities at university. *Active Learning in Higher Education*, *23*(2), 81–96. https://doi.org/10.1177/1469787420908209

The Hidden Meanings of 'Independent Research(er)'

Owen Gower

Doctoral research has traditionally been associated with the development of 'autonomy' and becoming an 'independent researcher' (see, for example, Johnson et al., 2000, p. 140). In this chapter, I suggest that the meaning of 'independence' in doctoral research is, in fact, ambiguous and contested. If I am right that 'independence' *is* ambiguous, then that might explain why it is part of the ***hidden curriculum*** in doctoral education. Policies, handbooks, and regulatory guidance all insist that researcher independence is an official goal of doctoral education. I contend, however, that ***unofficially*** there is considerable variation in the interpretation and importance placed on independence.

I explore the ways in which this ambiguity may affect doctoral recruitment, the identity of the researcher, the supervisory relationship, and doctoral assessment. Showing that, in some instances, a lack of clarity can be harmful, I evaluate whether a definition of 'researcher independence' could be standardised, made explicit, and therefore removed from the

O. Gower (✉)
UK Council for Graduate Education, Lichfield, UK
e-mail: Owen.Gower@ukcge.ac.uk

D. L. Elliot et al. (eds.), *Developing Researcher Independence Through the Hidden Curriculum*,
https://doi.org/10.1007/978-3-031-42875-3_21

hidden curriculum in doctoral education altogether. Fearing this to be an unlikely prospect, I conclude that 'independence' must be re-negotiated afresh by every doctoral researcher.

The *UK Council for Graduate Education* runs regular consultations with research supervisors to understand their attitudes and experiences of doctoral research. During a recent series of focus groups, conducted on behalf of UKRI, we asked whether the aim of doctoral education should be to produce 'independent researchers'. Surprisingly, there was disagreement, with some participants preferring instead the idea that research supervisors are aiming to inculcate 'research leadership' in their doctoral candidates. For some, 'independence' did not sufficiently capture the relational and collaborative aspects of research. Others were convinced, though, that the autonomy and agency of the doctoral researcher were rightly protected by characterising 'independence' as the goal of doctoral research. One participant pointed out, however, that the goal of doctoral education was, in part, determined by the doctoral researchers themselves, and for that reason could not be specified in advance. The implication is that those who are not intending to pursue an academic career may have no interest in becoming an 'independent researcher' in the way that that might be understood in university settings.

Scratch beneath the surface, then, and we find that there is disagreement about whether doctoral research should even involve becoming independent. Then there is the further complication that independence may mean different things to different people or in different contexts. Even the regulatory documents imply different connotations: one talks of 'independent *study*' (Quality Assurance Agency, 2020, p. 4); the other of 'independent *thought*' (Quality Assurance Agency, 2018, p. 4). I suggest that this kind of slippage tells us something about the wider, implicit, and sometimes contrasting meanings of 'independence' in doctoral research.

For example, 'independence', as used in 'independent *study*', implies that research methodology is important: that the doctorate has to be your own work, with minimal outside help, support, or interference, and best evaluated by arms-length surveillance. Whereas, 'independent *thought*' suggests something about the researcher themselves, better applied to 'a thinker' than a 'research outcome', and best evaluated in contrast with other thoughts and thinkers.

Does it matter that 'independent research(er)' might have different connotations? Well, 'yes' and 'no'. On one hand, the diversity of disciplines and doctoral programmes might naturally lead to differences in what counts as 'independence', and how much one ought to expect, and at what stage in the doctoral 'lifecycle'. What is more, there may be a reason why there is no standardised checklist for how to become an 'independent researcher': you might think that to achieve 'independence' one has to create the checklist for oneself.

On the other hand, if we do not have agreement on what 'independence' means, and how important it is to doctoral research, then clearly there is a possibility for misalignment between researchers, supervisors, and institutions. How we interpret 'independence' and whether it applies to the research or the researcher clearly has important implications for how candidates experience their doctorate. It also influences how we provide, describe, and assess doctoral research. Here are some areas where things might go wrong:

- for recruitment and selection—are we 'recruiting' independent researchers or 'developing' them?
- for the identity of the researcher—are they allowed to ask for help?
- for supervisory practice—what is the right balance between directive vs non-directive support?
- and for assessment—is the doctoral thesis 'being all your own work' more important than showing it to be distinct, or 'independent from other scholarship in the field'? And ultimately, how much does independence matter in doctoral examinations as compared to other doctoral characteristics?

Even at the application stage of doctoral research, 'independence' is given different weighting. Some doctoral applicants are expected to write research proposals with minimal support from potential supervisors. Others are recruited for doctoral projects conceived by research supervisors. And even when it is the norm for potential supervisors to write applications, that role can be resented. In recent research conducted by the *UK Council for Graduate Education*, for example, one respondent said:

> This is a particularly onerous and occasionally soul-destroying part of my job, as I have to come up with a high volume of projects for these students. (UK Council for Graduate Education, 2021)

Another suggested that the application process has become *less independent* in recent years:

> I'm spending more time writing PhD applications, [...] it's getting refined by potential supervisors and I spend quite a lot of time with some students writing it before the university even knows about it. (UK Council for Graduate Education, 2021)

Once enrolled, doctoral research is unlike other levels of study: there are fewer milestones; as a threshold qualification, there are no 'grades'; and there is significantly less scaffolding. The first experience of independent research is often not liberating but upsetting:

> You're not part of the staff but you're not like an undergraduate who goes in and does what they're told. So, you're in this kind of [...] this can be quite unsettling and it can influence and disrupt your work as well (part-time PhD, white man). (Morris, 2021, p. 7)

Some commentators have suggested that providers of doctoral education have even normalised the idea that developing independence necessarily involves psychological distress:

> The experience of isolation and abjection often appears so widespread as to be structural and endemic, a seemingly 'necessary' feature of the doctoral programme for many, rather than an accidental and ameliorable problem. Indeed, it may in some senses be a condition of the production of independence and autonomy, which is the goal of the pedagogy and practice of the PhD. (Johnson et al., 2000, p. 2)

In other words, 'independence' can be understood as 'being left alone to get on with it' and that can come at a significant psychological cost. Being 'unsettled', lonely, or isolated can be seen as the *price you have to pay to become an independent researcher*, and if you ask for help, you are not becoming one. This is clearly a toxic interpretation of 'independence', and Johnson et al. (2000) may be right to suggest that it derives from a 'masculine' conception of academic success. It is certainly the cause of much of the mental health concern we have for our doctoral researchers. What's more, it smacks of an outdated model of the 'lone researcher', when we know that modern research is increasingly driven by collaboration, co-creation, and 'team science'.

Nevertheless, cutting-edge research does involve a certain amount of 'figuring it out *for yourself*' by virtue of the fact that—by definition—no one has conducted *this* research in quite *this* way before. For that reason, independence is prized by research supervisors, some of whom are clearly concerned that the system already does 'too much' to support doctoral researchers:

> ...occasionally, students make it through the selection process and early evaluations (e.g. upgrading) because *too much* support is given to them (e.g. in developing their ideas/ help with writing etc). This is extremely dispiriting when the expectation is that everyone must pass their PhD and that it is a stain on your reputation as a supervisor if one of your students fails (even when they should). (UK Council for Graduate Education, 2021)

> Developments in higher education (taught programmes and research degrees) seem to have encouraged instrumentalism, superficiality and a reluctance among PGRs to develop the appropriate independence. (UK Council for Graduate Education, 2021)

So there is an inherent tension in 'independence': there can be both too little (if the supervisor is 'developing the ideas') and too much (if the doctoral researcher feels isolated), and no doubt this tension plays out afresh in every supervisory relationship.

Ultimately, how important is *independence* in doctoral research? Well, that is a contested question, even in the doctoral examination. It *is* a sought-after quality, but the doctorate is a 'threshold' qualification, and examiners are therefore entitled to some leeway in the relative importance they accord it:

> At the top end you're looking to say `Is this person going to make a competent and capable, independent...scientist?'...Would you be happy for this person to run a lab...do they have the...intellect, the rigour...do they have the integrity?' But there's a...let-out clause: this person may not be an independent scientist but have they completed a training period? Do they have enough skills to warrant the award of a title after their name? (STEM-E3). (Houston, 2018, p. 184)

Do any of these ambiguities undermine 'independence' as *a* goal of doctoral research? I don't think so, although I have grave concerns about how 'independence' and 'isolation' have become intertwined. Nevertheless, the formulation of research questions, the ability to seek out 'knowledge gaps', and a willingness to innovate and to take intellectual risks—all these we want to see in doctoral research, and all require a certain amount of independent thought. But we have also seen that 'independence' is a Goldilocks concept, and you can have too much and too little. So negotiating and interpreting the meanings of independence are inevitably part of each doctoral researcher's very own ***hidden curriculum***. We are left, in other words, in a situation where 'independence' is both the official goal of doctoral education, and yet resists a formulaic articulation which would suit *all* doctoral researchers and find agreement with *all* doctoral supervisors and examiners.

Discovering what independent research means for each doctoral researcher therefore requires a supportive research culture which tolerates different needs, attitudes, and expectations; a wealth of research support staff who can help the researcher recognise that they are achieving independence and agency on their own terms; and—perhaps most importantly of all—flexible and nuanced supervisory practices. But we would do a disservice to doctoral researchers to claim that the concept of an 'independent researcher' is never contested, or that it can be insulated from wider debates about the purpose, nature, and provision of doctoral education.

References

Houston, G. (2018). *A study of the PhD examination: Process, attributes and outcomes*. DPhil Thesis, University of Oxford.

Johnson, L., Lee, A., & Green, B. (2000). The PhD and the autonomous self: Gender, rationality and postgraduate pedagogy. *Studies in Higher Education, 25*(2), 135–147. https://doi.org/10.1080/713696141

Morris, C. (2021). "Peering through the window looking in": Postgraduate experiences of non-belonging and belonging in relation to mental health and wellbeing. *Studies in Graduate and Postdoctoral Education*.

Quality Assurance Agency. (2018). *UK quality code for higher education, advice and guidance for research degrees*.

Quality Assurance Agency. (2020). *Characteristics statement—Doctoral degree*.

UK Research Supervision Survey. (2021). UK Council for Graduate Education

Preparing for the World Outside Academia: Avoiding Organisational Culture Shock

Isabelle Skakni and Kelsey Inouye

Preparing for the World Outside Academia: Avoiding Organisational Culture Shock

In today's labour markets, individuals must adapt to increasingly dynamic work environments and more complex career trajectories. Academia is no exception. Over the past few decades, the growing number of PhD holders has exacerbated competition for tenure-track positions, and it is now widely acknowledged that most PhD holders will not stay in academia (OECD, 2021). Therefore, the hidden curriculum of doctoral education now includes the need for doctoral scholars to anticipate and prepare for a range of career paths.

I. Skakni (✉)
Research and Innovation Division, University of Applied Sciences and Arts Western Switzerland, Delemont, Switzerland
e-mail: Isabelle.Skakni@hes-so.ch

K. Inouye
Department of Education, University of Oxford, Oxford, UK
e-mail: kelsey.inouye@education.ox.ac.uk

D. L. Elliot et al. (eds.), *Developing Researcher Independence Through the Hidden Curriculum*,
https://doi.org/10.1007/978-3-031-42875-3_22

This chapter draws on the experiences of PhD holders pursuing careers beyond academia in Switzerland and the UK to explore the challenges of transitioning from academia to non-academic workplaces, a phenomenon we examined through the concept of organisational culture shock. The chapter highlights common sources of organisational culture shock—daily functioning of the workplace, organisational values, and statuses in the organisation—and discusses practical implications for doctoral scholars, universities, and those working in the researcher development space.

Integrating Non-academic Workplaces: Organisational Culture Shock

When PhD holders transition from their doctoral programmes or academic positions to roles in other sectors, they are moving into workplaces with distinctly different organisational cultures. Organisational cultures, understood as the shared beliefs, values, and attitudes held by an organisation's members (Alvesson, 2012), include the hierarchical structures, day-to-day practices, and expectations governing work, which new hires are meant to learn and adopt as part of the organisation (Ashforth et al., 2008). In universities, organisational cultures tend to be significantly distinct from those of organisations in other sectors. For instance, academics generally have high levels of autonomy in terms of the research they conduct, the courses they teach, and how they organise their time. Typically hired as experts, they also tend to identify more with their discipline and profession—academic researcher—than with their university or department (McMurray & Scott, 2013).

During their time in academic departments and research institutes, doctoral scholars are socialised into this unique organisational culture, including approaches to work, modes of interaction that are taken for granted, and the obligations and privileges associated with each status (e.g., professor, postdoc, doctoral scholar). Through what Gardner and Doore (2020) define as a process of professionalisation, doctoral scholars develop a sense of identity as members of the research profession in their field, internalising its values and norms and displaying them through their approaches to work, attitudes, and behaviour. After spending many years in academia, it may be challenging to integrate into another employment sector. In some cases, this difficulty translates into a puzzling experience in which one suddenly realises that others do not share their understanding of their work environment, resulting in what we define as an organisational culture shock.

Organisational culture shock is an individual's reaction to an unfamiliar environment that occurs when people are confronted with ways of working and thinking that differ from or contradict their own (Ward et al., 2005; Skakni et al., 2021). Organisational culture shock can range from surprise or astonishment at aspects of the new organisational culture to an inability to accept or comply with certain rules, norms, or expectations.

The Study

In this chapter, we report on interviews conducted with 16 PhD holders in Switzerland (n = 10) and the UK (n = 6), who experienced organisational culture shocks when transitioning from academia to other employment sectors. Table 1 presents their characteristics. The sample is comprised of PhD holders who were pursuing non-academic careers in private,

Table 1 Sample characteristics

Pseudo	*Gender*	*Age*	*PhD field*	*Years from graduation*	*Work sector*	*Work experiences before PhD*
UK						
Frances	F	36–40	HSS	10 years	Private	No
Felicity	F	36–40	HSS	7 years	Public	no
Adrian	M	31–35	HSS	6 years	Public	no
Amy	F	26–30	HSS	4 years	Public	no
Brenda	F	26–30	HSS	2 years	Public	no
Jim	M	36–40	HSS	Less than a year	Public	yes
Switzerland						
Jeremy	M	36–40	STEM	8 years	Private	no
James	M	41–45	HSS	6 years	Public	yes
Béatrice	F	36–40	HSS	6 years	Para-public	no
Helena	F	36–40	STEM	6 years	Private	no
Timothy	M	31–35	STEM	5 years	Private	no
William	M	36–40	HSS	4 years	Para-public/ High ed[a]	no
Alexander	M	31–35	HSS	2 years	Public	no
Alicia	F	36–40	HSS	1 year	Para-public/ High ed[a]	yes
Elizabeth	F	26–30	HSS	1 year	Para-public/ High ed[a]	yes
Colin	M	26–30	HSS	Less than a year	Private	no

[a]Part-time contract as lecturer

public, or parapublic sectors in Switzerland or the UK for ten years or less. The interviews, which took place online, explored how they found their current positions, the nature of their daily work, their PhDs' relevance and usefulness, and satisfaction with their current positions.

Participants' Organisational Culture Shock Experiences

Our findings show that when integrating into non-academic workplaces, most participants spent a significant part of their time and energy understanding their new workplaces' daily functioning, the values shared in their organisations, and the statuses they were assigned. This destabilising experience was more frequent in those who entered non-academic workplaces directly after their PhDs and those with little or no work experience prior to their PhD. In the following, we present examples of organisational culture shock these participants experienced.

Daily Functioning of the Workplace

The most common source of organisational culture shock for the participants (n = 13) was rooted in their new workplaces' daily functioning. Specifically, time management and scheduling, expectations for task performance, and modes of collaboration were key points of divergence between academic and non-academic organisational cultures, as most participants found themselves in highly structured environments compared to academia. For example, unlike in the PhD, where doctoral scholars are allowed substantial freedom to organise their time, work hours and timelines in their new organisations were rigid, with strict deadlines. Understanding the hierarchy was also a challenge.

> To adapt to this job, I need to develop the ability to interact well with the hierarchy, to do diplomacy and politics … to adapt to the schedules because the system is so rigid [….] This hierarchy stuff is very new to me. We always have to ask managers for their approval as if we couldn't think for ourselves! (Participant #8, Parapublic, Switzerland)

Participants who were used to being meticulous when conducting research in an academic setting found their organisations placed more emphasis on productivity than the depth of knowledge, requiring them to reframe how they approached writing and research tasks. Moreover, for several participants, collaborative work in their new organisations differed significantly from their previous academic experiences and was a challenge as they were expected to work collaboratively on projects with diverse viewpoints or ways of working.

> During my PhD, I worked a lot on my own. I was responsible for everything, from A to Z, and every now and again, I would check in with my supervisor, but I didn't really work with anybody, like in collaboration. So, that's something that—you never work on your own in a company. I'm still a lot more productive on my own than in a team, but that's not very productive for the company because a lot of the time, the job requires you to work with many other people from many other teams (Participant #14, Public, UK)

Organisation's Values
For some participants (n = 6), organisational culture shock manifested in relation to their new organisation's values, which speak to what they consider important in their work and work conditions. For some, entering a non-academic workplace triggered reflections on the purpose they attach to work in general. For instance, some participants missed the passion for research work they harboured in their academic careers, as illustrated in the following:

> [There is a] lack of passion. I mean, research is a lifestyle, is a.... It monopolises your life. It's like a lover—I don't know how to say. So, this [current position] is a job. It's not a mission. I miss it a lot, I must say. (Participant #11, Private, UK)

Others highlighted a clash between the degree of personal commitment expected in academia and the norms in their non-academic workplace. In some cases, expectations from their organisations were greater than what they had experienced in academia (e.g., long work hours, travelling abroad every week, etc.).

Statuses in the Organisation
Participants (n = 5) also described organisational culture shock experiences related to their organisational statuses. These challenges refer to their perception of the duties, obligations, and entitlements that come with their positions. For instance, some participants had trouble determining their place in the organisation; as doctoral scholars or academics, they were treated as experts in their fields. In their new organisations, where the nature and value of a PhD were not necessarily understood or recognised, having to re-establish themselves and 'prove' their expertise could be frustrating.

> Only 20% of my knowledge is used here. It's my biggest regret; I developed a particular knowledge over the years. I was an expert, and overnight all went up in smoke. I literally threw it away [by taking this position]. (Participant #1, Private, Switzerland)

In some cases, having an academic background confers a blurred status, affecting interaction with colleagues.

> When I arrived as a psychologist with a PhD, I didn't feel welcome; I was seen as a bit of a threat. I didn't think that would be a problem at all. But I realised that it was better to keep a low profile if I wanted to fit in. (Participant #8, Parapublic, Switzerland)

Overall, the concept of organisational culture shock highlights that beyond developing skills that can be mobilised in a range of positions, doctoral scholars would also benefit from early exposure to the expectations and work traditions in non-academic sectors—a crucial component of the hidden curriculum of doctoral education.

Avoiding Organisational Culture Shock: Insights for Policy and Practice

As revealed in this study, the transition from academia to other sectors can prove challenging for some PhD holders due to the differences in organisational cultures, manifested primarily in the daily functioning of the workplace, the organisation's values, and statuses in the organisation. Preparing for the post-PhD period is a fundamental part of the hidden curriculum of doctoral education because it has become necessary for doctoral scholars to build on the knowledge and skills developed during their doctorates to steer their careers according to their personal affinities and aspirations. It is therefore crucial to support doctoral scholars in preparing for various types of careers, equipping them for the practicalities of working outside academia, and fostering independence in career planning and skill mobilisation. (This argument can be linked to the rationale that underpins Aarnikoivu's, Dai's and Dangeni et al.'s chapters.) In the following, we highlight our study's practical implications for current doctoral scholars and the main stakeholders involved in facilitating early-career researchers' development.

Doctoral Scholars

As early as possible in their doctoral journeys, doctoral scholars should start exploring employment sectors and types of jobs they would like to hold. An effective strategy is for doctoral scholars to identify the tasks or aspects of their doctoral experience they find particularly enjoyable and explore work sectors or types of positions that involve related tasks and aspects. It can also be helpful to determine what matters most in their work (e.g., Do they value collaboration? Teaching and mentorship? The research itself?), the lifestyle that suits them best, and the values that are important to them in the context of work. They should keep these concerns in mind when exploring jobs.

Moreover, being proactive in learning about the range of careers available by drawing on networks, attending seminars, or participating in placements and similar programmes may be a means of empowerment, allowing doctoral scholars to identify possibilities and ascertain the types of work environments and organisational cultures associated with various career options. It can also be valuable to talk with PhD holders in their field who have already left academia to find out how they experienced this transition, what challenges they encountered, and what strategies were most successful. Similarly, having a career-focused conversation with a doctoral supervisor, academic line manager, or researcher developer may be helpful in defining career goals and identifying resources for support. In these ways, interdependence amongst doctoral scholars and support from academic staff may be central to gaining insight into the realities of career transitions and how to navigate them. However, this commitment of doctoral scholars to prepare for various types of careers cannot be achieved without adequate institutional support.

Universities, Researcher Developers, and Supervisors

It is crucial for universities to acknowledge that supporting doctoral scholars to prepare for various careers entails culture change within academia. Universities must proactively develop and promote opportunities to develop a range of skills and expertise, thus recognising and normalising increasingly diverse career trajectories. A first step is to provide PhD holders with precise and up-to-date information about the range of career opportunities available within and outside academia, recognising that options differ across disciplinary fields. Being aware of actual opportunities may help doctoral scholars explore and prepare for various types of work while normalising the idea of the PhD as preparation not only for a specialised knowledge area but also for any career requiring high-level critical thinking and communication skills. In the same vein, it is also necessary to make thesis supervisors aware of the importance of encouraging their supervisees to explore career options and critically reflect on their career aspirations.

Career planning, including an individualised career plan that considers doctoral scholars' background, strengths, and interests, should be a formal part of PhD programmes. Universities should also offer courses or seminars that enable doctoral scholars to communicate with various audiences, which may help them engage non-academic sectors in their research and provide experience in writing in various genres, a skill that is valued in organisations across sectors. Similarly, doctoral scholars should have the opportunity to participate in internships, policy engagement programmes, and industry partnerships that allow them to experience work cultures in other organisations and facilitate networking in various sectors. In this regard, career developers and, more broadly, universities' career centres, would benefit from collaborating closely with stakeholders from non-academic sectors, enabling them to understand better how these employment sectors are organised, what their recruitment traditions and expectations are, and what career opportunities they concretely represent for PhD graduates. Effective PhD career preparation means helping doctoral scholars to develop flexibility in both their skill sets and understandings of what their career options can be, requiring interdependence with peers in their research community, i.e., a network of individuals including the doctoral scholar, supervisors, and support from the universities themselves.

Conclusion

This chapter addressed the preparedness of doctoral scholars for diverse career paths in a context where academic positions are increasingly difficult to obtain. Drawing on the concept of organisational culture shock, we argued that anticipating and preparing for different types of careers is now a core part of the hidden curriculum of doctoral education. This is particularly critical to the development of researcher independence, as awareness of career options and how to mobilise one's skills empowers doctoral scholars to shape their futures in whatever sectors they choose. At the same time, building researcher independence requires institutional and supervisor support. Fostering interdependence amongst doctoral scholars, PhD graduates, and academic and professional staff is essential in creating a community of resources and knowledge in which doctoral scholars can explore their career prospects and engage in professional development opportunities—potentially mitigating organisational culture shock and providing doctoral scholars with greater confidence in their career trajectories.

Declarations of Conflicts of Interest No potential conflict of interest is reported by the authors.

Funding This work was supported by an Erasmus+ programme of the European Union under the project 'Researcher Identity Development: Strengthening Science in Society Strategies' [grant number 2017-1-ES01-KA203-038303].

References

Alvesson, M. (2012). *Understanding organizational culture.* Sage.

Ashforth, B. E., Harrison, S. H., & Corley, K. G. (2008). Identification in organizations: An examination of four fundamental questions. *Journal of Management, 34*(3), 325–374. https://doi.org/10.1177/0149206308316O

Gardner, S. K., & Doore, S. A. (2020). Doctoral student socialization and professional pathways. In J. C. Weidman & L. DeAngelo (Eds.), *Socialization in higher education and the early career: Theory, research and application* (pp. 113–127). Springer.

McMurray, A., & Scott, D. (2013). Determinants of organisational climate for academia. *Higher Education Research & Development, 32*(6), 960–974. https://doi.org/10.1080/07294360.2013.806446

OECD. (2021). *Reducing the precarity of academic research careers.* OECD Science, Technology and Industry Policy Papers, 113. OECD Publishing.

Skakni, I., Inouye, K., & McAlpine, L. (2021). PhD holders entering non-academic workplaces: Organisational culture shock. *Studies in Higher Education, 47*(6), 1271–1283. https://doi.org/10.1080/03075079.2021.1876650

Ward, C., Furnham, A., & Bochner, S. (2005). *The psychology of culture shock.* Routledge.

Changing Career Pathways: Making Visible the Employment Destinations and Non-academic Contributions of Doctoral Scholars

Sally Hancock

Introduction

This chapter examines the changing employment outcomes of doctoral scholars. The continued expansion of doctoral education in recent decades has meant that most doctoral scholars will undertake non-academic employment after completing their programme (Cyranoski et al., 2011; Hayter & Parker, 2019; Sauermann & Roach, 2012). Although labour markets vary by national context, the tendency towards increased competition for academic positions is observed in most research systems, including the United Kingdom, which serves as the empirical case study for this chapter.

S. Hancock (✉)
University of York, York, UK
e-mail: sally.hancock@york.ac.uk

D. L. Elliot et al. (eds.), *Developing Researcher Independence Through the Hidden Curriculum*,
https://doi.org/10.1007/978-3-031-42875-3_23

The topic of doctoral expansion and changing employment outcomes is polarising (Hancock, 2021). Those committed to the project of the knowledge-based economy welcome this as evidence that doctoral scholars' high skills and expertise are readily absorbed into the non-academic labour market. In countries such as Germany, where the notion of a dual-purpose PhD is well established, the idea of preparing doctoral scholars for careers beyond academia is less contentious. However, in the United Kingdom, the economic reframing of the doctorate is the result of a concerted policy effort in recent years.

UK governments have invested considerably in doctoral funding, skills training, and programme reform to support doctoral scholars' progression into the non-academic labour market. Early initiatives focused on science, technology, engineering, and mathematics (STEM) but doctoral scholars across all fields are now encouraged to undertake transferable skills and professional development training and to actively prepare for diverse career outcomes (Roberts, 2002; ESRC 2021; EPSRC 2021). At the time of writing, UK Research and Innovation, the major public funder of doctoral study, is consulting on a 'new deal' for postgraduate research which aims to secure inclusive and diverse career pathways for doctoral scholars (UKRI, 2022).

In contrast to the optimism of political narrative, a growing body of literature documents that doctoral scholars often feel unprepared for and discontented with these developments. The aspiration to secure an academic research career remains a priority for many. Non-academic opportunities may be considered late in the doctorate or even postdoctoral stage. Doctoral supervisors, disciplinary norms, and institutional cultures are understood to reinforce a preference for academic research careers and the depiction of non-academic employment as a secondary, less well-aligned career outcome (Hayter & Parker, 2019; McAlpine & Amundsen, 2018; Sauermann & Roach, 2012; Suomi et al., 2020). Furthermore, the prospect of uncertain career outcomes is a significant source of poor mental health and well-being among doctoral scholars (Walsh & Juniper, 2009; Tazzyman et al., 2021).

This chapter aspires to illuminate two 'hidden aspects' of contemporary doctoral education: the employment destinations of doctoral scholars and the value of doctoral knowledge, skills, and experience to non-academic employers. It does so through an analysis of large-scale employment data on doctoral scholars in the United Kingdom. The chapter pays particular attention to the non-academic employment roles that the majority of doctoral scholars go on to undertake, and explores the hidden learning—the informal knowledge, skills, and experiences—that non-academic employers value. Doctoral scholars are encouraged to reflect on the many pathways open to them after the PhD, and on the importance of developing these aspects of hidden learning alongside the formal components of their doctorate. It is argued that doing so will foster the independence of doctoral scholars as they prepare to embark on diverse career pathways. (This sentiment is aligned with Skakni and Inouye's chapter advocating enhancing doctoral scholars' readiness for increasingly diverse career paths.)

Independence here is understood through a sociological lens that considers the interplay between agency and structure and the ways in which individual efforts, aspirations, and values are both enabled and constrained by other actors, organisations, and structures (Hodkinson & Sparkes, 1997; McAlpine et al., 2014). As noted in previous work on doctoral scholars' career planning and decision-making, independence is an ongoing, reflective process which involves an awareness of individual values, knowledge, and skills, as well as relevant policies and economic factors that shape employment outcomes (Hancock et al., 2017; Hancock, 2019).

Through its exploration of career destinations and the value of doctoral knowledge, skills, and experience beyond academia, the chapter encounters previously unrecognised aspects of the hidden curriculum: namely, the development of doctoral *skills and employability* through unofficial mechanisms of learning (Elliot et al., 2020). The chapter is structured as follows. The methods of the study are described in the next section, followed by an overview of empirical observations of changing doctoral careers in the UK. The chapter concludes with a discussion of these findings, reflecting on implications for the conceptualisation of the hidden curriculum in doctoral education, and drawing out recommendations and actions for doctoral scholars and other stakeholders.

Methods

The findings presented in this chapter arise from a wider study of doctoral employment in the United Kingdom. The study made use of secondary data on doctoral employment in the UK, an approach that has been adopted in other contexts (Li & Horta, 2022; Skovgaard Pedersen, 2014). Unlike countries including the United States and Germany, the UK does not systematically track the careers of doctoral graduates. The best large-scale evidence on doctoral employment available in the UK is found through sector-wide graduate surveys (formerly, the Destinations of Leavers from Higher Education; now, Graduate Outcomes).

The chapter draws from the longitudinal Destinations of Leavers from Higher Education survey (henceforth, Long DLHE), which captures employment activity three and a half years after graduation. The most recent Long DLHE data were provided by the Higher Education Statistics Agency (HESA) at the time of undertaking the research. This resulted in a sample of 4731 UK-domiciled PhD graduates who completed their degrees in 2008/2009 and 2010/2011 (response rate: 39.5%). This represents around one-fifth of the UK-domiciled PhD population graduating in those years. Data were weighted by HESA prior to analysis to reflect academic variables. All analyses were conducted in Stata and reporting adheres to the HESA Standard Rounding Methodology, which aim to prevent the identification of individuals. Table 1 details the distribution of academic and demographic characteristics in the sample.

Table 1 Participant demographics

Survey year	
2011/2012	46.7
2013/2014	53.3
Doctoral institution	
Russell Group	61.7
Other	38.3
Doctoral subject	
Arts and humanities	14.8
Biological sciences	21.2
Biomedical sciences	16.9
Physical sciences and engineering	32
Social sciences (including education)	15.1
Entry qualification	
Undergraduate qualification only	54.2
Taught master's degree	45.8
Gender	
Male	49.7
Female	50.4
Ethnicity	
White	90.4
Asian	5.3
Black	1.2
Other (including mixed)	3.1
Age	
Under 30	49.3
30 and over	50.7
Parental home	
Low participation neighbourhood	7.6
Other neighbourhood	92.4

Note: All figures are percentages ($n = 4288$)

Changing Career Pathways: Observations from the United Kingdom

The long DLHE dataset indicates many positive employment outcomes for doctoral scholars. Unemployment rates are low, with the majority engaged in paid work or further training three and half years after completion (97.8%). The mean salary reported by doctoral graduates (£44,917) is higher than the current national average. Career satisfaction is also high (91.8% reported being very or fairly satisfied with their circumstances). Nevertheless, the prospect of leaving academia is the likely outcome for most doctoral scholars. Employment pathways outside of academia are diverse and subject-specific and highlight the importance of developing additional skills and experience alongside the doctoral qualification. It is on these aspects that the following section will focus.

Employment Within and Beyond Academia

While most doctoral scholars (70.1%) leave the academic sector some three and a half years after completing their programme, departure rates vary markedly by subject area. This is demonstrated in Fig. 1.

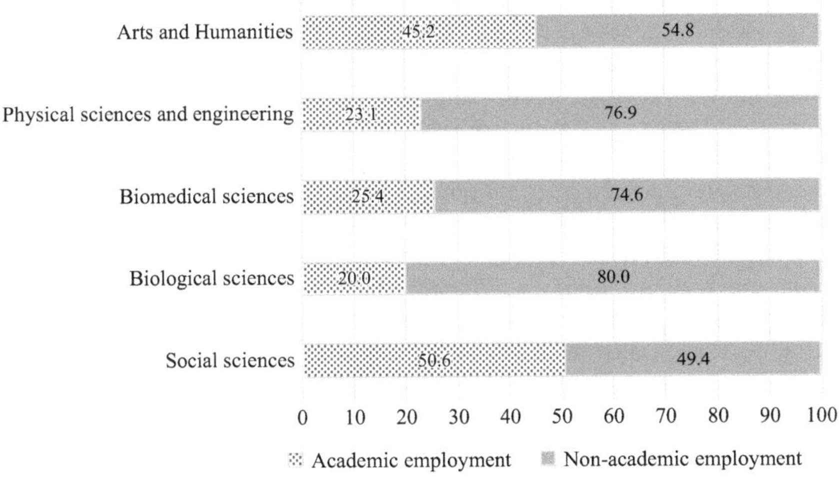

Fig. 1 Academic and non-academic employment by doctoral subject area. (Note: All figures are percentages (n = 4288))

To an extent, the differences observed between subject areas reflect the very different numbers of doctorates awarded in distinct fields. The majority of UK doctorates are awarded in scientific fields (HESA, 2022), and it is to be expected that here the competition for academic posts is most crowded. Nevertheless, beyond academia, it appears that there is a high demand for the research expertise and skills of doctoral scholars in scientific fields.

Working Beyond Academia: Research and Non-research Roles

Since most doctoral scholars are likely to enter non-academic employment, it is pertinent to explore the types of non-academic roles occupied in further detail. As the PhD is a research degree intended to develop researcher independence, it is instructive to assess the extent to which non-academic roles are perceived to incorporate an element of research. In this analysis, a research role is characterised as one where the creation, application, or dissemination of research is the duty of the post holder.

The categories of research and non-research roles were developed because there is currently no agreed definition of a doctoral job, in the way that researchers share consensus over the nature of 'graduate jobs' (Elias & Purcell, 2013). To be coded as occupying a non-academic 'research' role, several criteria needed to be met. First, individual occupational titles were assessed. In some cases, a connection to research was clear (for example, 'Biological scientists and biochemists'), whereas others, such as a primary school teacher, could be discounted. Second, all cases with occupational titles suggestive of research were assessed against information about the formal requirements of employment. To remain coded as a research role holder, a doctoral scholar must have reported that the qualification, subject knowledge, skills and competencies, *or* practical experience and work placements gained during the doctorate were a formal requirement of employment. This is indicative that the research role was positioned at the doctoral level. For doctoral scholars who leave the academic sector, employment in research and non-research roles is approximately equal.

Slightly over half (53.8%) of those employed outside of academia report holding a research role; meaning that some 46.1% of those working beyond academia are not employed in a research-related capacity. Further

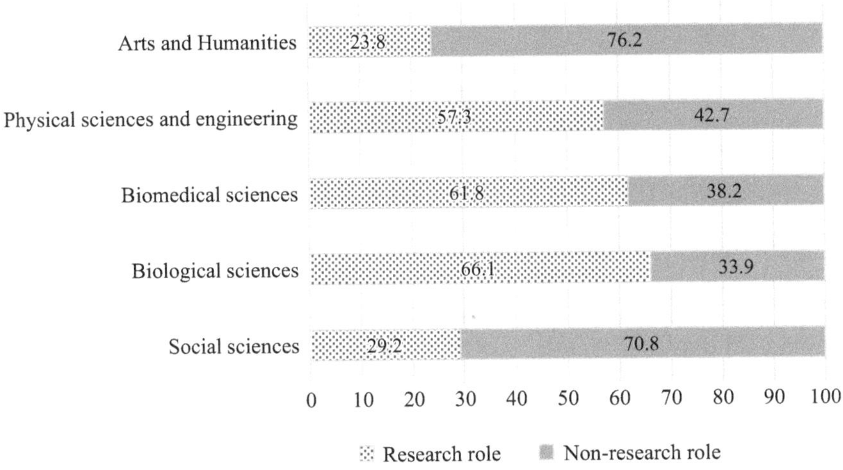

Fig. 2 Research and non-research roles in non-academic employment by doctoral subject area. (Note: All figures are percentages (n = 3007))

analysis by doctoral subject area indicates that routes in research employment beyond academia are highly field-specific. Figure 2 displays the coding of doctoral graduates in research and non-research roles outside of academia by subject area.

It is striking that though high numbers of science doctoral scholars leave academia, the majority go on to secure research employment beyond academia. This suggests that non-academic employers value the research training of science doctoral graduates highly, and that research-relevant employment opportunities are abundant in the UK labour market. Interpreting the outcomes for Arts and Humanities and Social Science doctoral scholars is somewhat more ambiguous. The lower respective rates of research employment for these graduates may reflect structural factors underpinning the demand for different types of doctoral scholars in the non-academic labour market. However, it may also reflect agentic differences in the extent to which doctoral scholars are positioned to articulate the relevance of their knowledge, skills, and experience to non-academic employers. The difference in these outcomes, and the implications for facilitating doctoral scholars' independence, is considered further in the discussion.

Role Requirements: Knowledge, Skills, and Experience

Doctoral scholars also shared information about the formal requirements of their employment. Figure 3 shows this data comparatively: for those occupying academic roles, non-academic research roles, and non-academic non-research roles.

The doctoral qualification matters most for academic employers (78.7% of doctoral scholars in these roles identified this as a formal requirement). It is also important for those employed in research roles beyond academia (56.8%), but notably less so for those occupying non-research roles (18.7%). Similarly, doctoral subject knowledge is sought most by academic (66.4%) and non-academic research employers (53.4%). Only one-quarter of those employed in non-research roles beyond academia identified this as a formal requirement of their post. Differences in formal role requirements lessen in relation to doctoral skills and competencies, and work experience gained since the doctorate: across roles and sectors, these are valued by upwards of half of all post holders. These data emphasise the importance of transferable skills and competencies and of continued learning beyond the doctoral degree. Work experience undertaken during the doctorate is less essential for academic and non-academic research

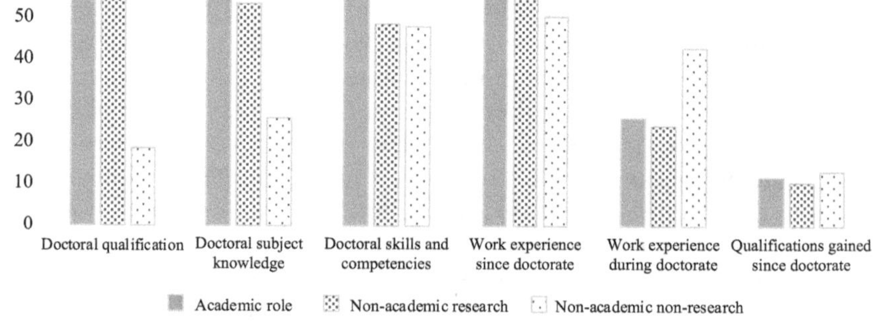

Fig. 3 Formal role requirements by academic and non-academic employment. (Note: All figures are percentages (n = 4288))

employers (only around one-quarter of post holders identified this as a formal requirement). However, almost half of those employed in non-research roles outside of academia related this to be a formal requirement.

Across all roles and sectors, qualifications gained since the doctorate are highlighted by only a minority. Reflecting on the variations in formal role requirements, it is notable that for most items, the agreement rate is lower for those working in non-research roles beyond academia. This may indicate that a broader set of specifications than those captured in the survey are valued by non-academic employers, a consideration explored in the subsequent discussion.

Discussion

Through its analysis of career destinations and the value of doctoral knowledge, skills, and experience beyond academia, this chapter has made manifest two hidden aspects of doctoral education. First is the visibility of doctoral employment destinations. Consistent with earlier studies of doctoral employment, the analysis of UK data indicates many positive outcomes for doctoral scholars. Though the majority can expect to leave academia, this should not necessarily be regarded as a concerning outcome: doctoral scholars secure diverse, well-paid, satisfying work. Nevertheless, there are differences in the type of work doctoral scholars do upon leaving academia. Pathways into research roles in particular appear to be field-dependent, with science doctorates more likely to secure such roles. The long DLHE, however, provides little insight into the causes of these differences.

Such distinctions likely reflect structural aspects of the UK economy but may also reveal differences in the ways in which doctoral scholars position themselves for potential, and particularly non-academic, employers. The second hidden aspect to emerge from this analysis is the importance of communicating the value of the doctorate to non-academic employers. Viewing the hidden curriculum from an economic perspective prompts a reappraisal of its conceptualisation in doctoral education. While earlier definitions emphasised academic, personal, social, and psychological needs, the analysis presented here suggests that the development of *skills and employability* also constitute a significant and valuable strand of hidden learning during the doctorate (Elliot et al., 2020).

Turning to consider implications for fostering independence, what can doctoral scholars and those supporting them learn from the experiences of those on non-academic trajectories? The data reveal many structural and demand-side aspects of doctoral employment that may helpfully inform career planning. First, while the doctorate is valued by non-academic employers, the qualification is not as essential outside academia as it is within it. Non-academic employers seek varied skills, competencies, and experiences, highlighting the importance of nurturing hidden learning during the doctoral programme. Studies from other national contexts support this observation. Drawing on an analysis of different European labour markets, Kyvik and Olsen (2012) concluded that 'generic skills' such as analytical thinking and problem-solving were most valued by non-academic employers. In the United States, Nerad et al. (2008) similarly reported that critical thinking, presentation skills, and group working were most sought by non-academic employers.

It is therefore essential that doctoral scholars permit time to develop a broad portfolio of skills, competencies, and professional experience during their programme. Here, scholars in the sciences may have an advantage since an emphasis on transferable skills development, collaborative working, internships, and the articulation of the non-academic value of research is long established (Roberts, 2002). It is not the case that doctoral scholars in the arts, humanities, and social sciences have less to offer to non-academic employers. However, professional development opportunities and career guidance have historically been less advanced in these disciplines (British Academy, 2020).

How, then, might such hidden learning for changing careers be harnessed to nurture the independence of all doctoral scholars? In the UK, resources such as the Researcher Development Framework and the Concordat provide a framework to support doctoral scholars and their supervisors to develop broad skills.[1,2] Internships and placements, such as those with the UK Parliamentary Office for Science and Technology, provide an opportunity to network and apply doctoral learning in a new setting. The commitment from two of the UK research councils to formalise placements and public engagement opportunities for all funded doctoral scholars suggests that the importance of once informal, hidden, and additional learning is now being recognised (EPSRC, 2021; ESRC, 2021). Scholar-led initiatives, for example, establishing writing groups, engaging

[1] https://www.vitae.ac.uk/researchers-professional-development/about-the-vitae-researcher-development-framework

[2] https://researcherdevelopmentconcordat.ac.uk/

with societies, networks, volunteering, and public outreach, will further support the development of a broad portfolio of skills, competencies, and professional experience (Elliot et al., 2020).

Developing an awareness of structural, economic, and policy realities is a critical aspect of fostering doctoral scholars' independence, but it is certainly not the only component. Individual motivations, agency, and decision-making are equally influential in shaping career outcomes; and doctoral scholars' aspirations are varied and fluid (Hancock et al., 2017; Hancock, 2019; McAlpine & Amundsen, 2018). The intention of this chapter is not to dismiss doctoral scholars' ambitions of becoming an academic, but to situate this outcome as one among many valuable pathways. Even within the academic sector, forming partnerships with government, private, and third-sector organisations and developing research impact are increasingly important aspects of academic practice. Time invested in the development and articulation of hidden skills and employability learning will therefore strengthen the career decision-making and agency of all doctoral scholars (Bryan & Guccione, 2018). To prioritise only the formal doctoral curriculum, with its focus on the development of a thesis and academic outputs, risks limiting the professional horizons and independence of all doctoral scholars (Elliot et al., 2020). A curriculum encompassing formal and hidden learning, tailored to individual aspirations and needs, will better align the contemporary doctorate with changing career pathways.

Declarations of Conflicts of Interest N/A

Acknowledgements The data referred to in this chapter were acquired with financial support from the Society for Research into Higher Education under Grant NR201609. The dataset was prepared by the Higher Education Statistics Agency of the UK and is not openly available.

References

British Academy. (2020). *Good practice in providing careers guidance to postgraduate research students.* https://www.thebritishacademy.ac.uk/publications/career-pathways-postgraduate-research-students-good-practice/

Bryan, B., & Guccione, K. (2018). Was it worth it? A qualitative exploration into graduate perceptions of doctoral value. *Higher Education Research & Development, 37*(6), 1124–1140.

Cyranoski, D., Gilbert, N., Ledford, H., Nayar, A., & Yahia, M. (2011). Education: The PhD factory. *Nature, 472*, 276–279.

Elias P., & Purcell, K. (2013). *Classifying graduate occupations for the knowledge society.* https://warwick.ac.uk/fac/soc/ier/futuretrack/findings/elias_purcell_soche_final.pdf

Elliot, D. L., Bengsten, S. S. E., Guccione, K., & Kobayashi, S. (2020). *The hidden curriculum in doctoral education.* Palgrave Macmillan.

EPSRC. (2021). *Review of EPSRC-funded doctoral education.* https://www.ukri.org/wp-content/uploads/2021/10/EPSRC-071021-ReviewDoctoralEducationSupport.pdf

ESRC. (2021). *Review of the PhD in the social sciences: ESRC response.* https://www.ukri.org/wp-content/uploads/2022/03/ESRC-110321-PhD-Review-Response.pdf

Hancock, S. (2019). A future in the knowledge economy? Analysing the career strategies of doctoral scientists through the principles of game theory. *Higher Education, 78*(1), 33–49.

Hancock, S. (2021). What is known about doctoral employment? Reflections from a UK study and directions for future research. *Journal of Higher Education Policy and Management, 43*(5), 520–536.

Hancock, S., Hughes, G., & Walsh, E. (2017). Purist or pragmatist? UK doctoral scientists' moral positions on the knowledge economy. *Studies in Higher Education, 42*(7), 1244–1258.

Hayter, C., & Parker, M. (2019). Factors that influence the transition of university postdocs to non-academic scientific careers: An exploratory study. *Research policy, 48*(3), 556–570.

HESA. (2022). *HE student enrolments by subject of study and domicile.* https://www.hesa.ac.uk/data-and-analysis/students/what-study

Hodkinson, P., & Sparkes, A. C. (1997). Careership: A sociological theory of career decision making. *British Journal of Sociology of Education, 18*, 29–44.

Kyvik, S., & Bruen Olsen, T. (2012). The relevance of doctoral training in different labour markets. *Journal of Education and Work, 25*(2), 205–224.

Li, H., & Horta, H. (2022). Factors influencing PhD students' intentions to pursue careers in the government and nonprofit sectors: Evidence from a global survey. *Higher Education Research & Development, 41*(6), 1946–1961.

McAlpine, L., & Amundsen, C. (2018). *Post-PhD career trajectories: Intentions, decision-making and life aspirations.* Palgrave Macmillan.

McAlpine, L., Amundsen, C., & Turner, G. (2014). Identity-trajectory: Reframing early career academic experience. *British Educational Research Journal, 40*, 952–969.

Nerad, M., Rudd, E., Morrison, E., & Picciano, J. (2008). *Social science PhDs—Five+ Years out: A national survey of PhDs in six fields.* https://www.education.uw.edu/cirge/phd-career-path-tracking/2261-2/

Roberts, G. (2002). *SET for success: The supply of people with science, technology, engineering and mathematics skills*. HM Treasury.

Sauermann, H., & Roach, M. (2012). Science PhD career preferences: Levels, changes, and advisor encouragement. *PLoS One, 7*(5), e36307.

Skovgaard Pedersen, H. (2014). New doctoral graduates in the knowledge economy: Trends and key issues. *Journal of Higher Education Policy and Management, 36*(6), 632–645.

Suomi, K., Kuoppakangas, P., Kivistö, J., Stenvall, J., & Pekkola, E. (2020). Exploring doctorate holders' perceptions of the non-academic labour market and reputational problems they relate to their employment. *Tertiary Education and Management, 26*, 397–414.

Tazzyman, S., Moreton, R., Bowes, L., Wakeling, P., & Stutz, A. (2021). *Review of the PhD in the Social Sciences*. https://www.ukri.org/publications/review-of-the-phd-in-the-social-sciences/

UKRI. (2022). *New deal for postgraduate research: Call for input*. https://engagementhub.ukri.org/skills/new-deal-for-postgraduate-research-call-for-input/user_uploads/new-deal-for-postgraduate-research-call-for-input%2D%2D-alternative-format-1.pdf

Walsh, E., & Juniper, B. (2009). *Development of an innovative well-being assessment for postgraduate researchers and its use to drive change*. Paper presented at the annual Vitae Researcher Development Conference, University of Warwick, September 8–9.

Concluding Thoughts

Dely Lazarte Elliot, Kay Guccione, and Søren S. E. Bengtsen

In deciding to embark on a follow-up project to our book *The Hidden Curriculum in Doctoral Education* published in 2020, we regard the expansion of conceptual understanding of the hidden curriculum as key to our mission while also exploring how lessons in different contexts can be harnessed. Given other concepts deemed crucial in doctoral education, we had agreed that 'twinning' the hidden curriculum with researcher independence was an appealing topic to explore—both conceptually and pragmatically. Along with related notions of agency, engagement and well-being, these concepts have been featured in each chapter, becoming the focal points for discussion. The contributions are from 45 scholars based in different geographical regions and different national, institutional and cultural contexts, and represent five doctoral stakeholder groups (doctoral scholars, supervisors, researcher developers, institutional leaders and external partners). They offer complementary views as well as challenging

D. L. Elliot (✉) • K. Guccione
University of Glasgow, Glasgow, UK
e-mail: Dely.Elliot@glasgow.ac.uk; Kay.Guccione@glasgow.ac.uk

S. S. E. Bengtsen
Aarhus University, Aarhus, Denmark
e-mail: ssbe@edu.au.dk

D. L. Elliot et al. (eds.), *Developing Researcher Independence Through the Hidden Curriculum*,
https://doi.org/10.1007/978-3-031-42875-3_24

contemporary understandings of how doctoral education may be further sustained. And so, this has enabled this edited collection to offer a much richer educational and institutional discourse and pedagogical framing.

So, what have we learned from this collective book project? Let us start by examining further the conceptualisation that underpins the hidden curriculum. Acquired collective wisdom made us reflect on what *hiddenness* actually entails. Does it perhaps refer to doctoral scholars not being aware of the hidden doctoral pedagogies because they are not customarily part of the doctoral structure? Or can this be explained by the idea that lessons are found at the periphery, and in turn, this easily leads to incidental types of learning? It stands to reason that when activities are not compulsory, doctoral participation could be 'hit and miss'. As a result, access to hidden curricular lessons may not only be variable; for some, they may remain undiscovered. In other words, the 'darkness' of the hidden curriculum may indeed hold both constructive and potentially worrying elements if left in a blind angle (Bengtsen & Barnett, 2017).

When doctoral scholars discover the hidden curriculum and tap into its benefits, is the search driven by these scholars themselves or guided by other hidden curriculum agents (e.g. supervisors, researcher developers, institutional leaders)? The answer to this is linked to the question—whose responsibility is it to facilitate 'access' and harness lessons emanating from the hidden curriculum?

To an extent, we argue that the lack of structure in the doctoral context leads to both opportunities and barriers. Specifically, a lack of structure can create ample opportunities for expanding pedagogical practices to support doctoral scholars' personal, professional and career development (Brodin & Avery, 2020; Elliot, 2022). Searching for such opportunities is typically driven by doctoral scholars themselves or facilitated by other doctoral and institutional stakeholders. On the contrary, the combined lack of awareness or inaccurate understanding of the role of the hidden curriculum can inadvertently work against these scholars and subsequently hinder their progress (Bengtsen, 2016). Understanding this suggests that doctoral scholars will reap the benefits should they become more aware of the 'what' (characteristics), 'how' (learning acquisition) and 'why' (value) of the hidden curriculum (Elliot et al., 2016). This leads nicely to discussing the varying *nature* and *degree of hiddenness* within the hidden curriculum, which often hinges upon scholars' personal efforts and support received.

The Hidden Curriculum as a Continuum

Due to the multiplicity of factors characterising the hidden curriculum, it is arguably best presented as a continuum, which ranges from the typically accessible through the not yet accessible or not yet existing doctoral pedagogies. We have identified three domains in the continuum of the hidden curriculum.

In Table 1, we theorise that hidden curricular learning comes from a continuum of pedagogies and resources that are available to doctoral

Table 1 Continuum of accessible and non-accessible hidden curricular pedagogies

	Structured pedagogies (required)	*Informal pedagogies (optional)*	*Not yet existing pedagogies (intentional)*
Aim	To learn beyond	To learn outside (e.g. academic, transferable skills)	To create learning (e.g. academic, transferable skills)
Driven by	Institutional programmes, strategies and culture	Personal and academic needs, career aspirations	Personal and academic needs, career aspirations
Sample activities	Supervisory meetings, seminars, workshops	Internships, clubs & societies, third spaces	Journal clubs, writing groups, creative supervision, innovative approaches to training
Viewed as	Obligatory or highly recommended	Requires extra effort	Requires co-creation or co-organisation
Approach	'Connecting the dots'	Enhancing experience	Creating new experience, community building
Tools	Reflection, self-evaluation metacognition	Reflection, self-evaluation proactive search	Reflection, proactive search creativity
Hidden curricular learning	Initiated by doctoral scholars	Initiated by doctoral scholars	Initiated by doctoral scholars, supported by supervisors, researcher developers or institutions
Added benefits	Personal, professional advancement	Personal, professional well-being advancement	Personal, professional, well-being advancement

scholars, i.e. structured pedagogies on the one hand, informal pedagogies in the middle, and pedagogies that are not yet currently existing on the other hand. Encompassing doctoral pedagogies (or genuine experiences of learning) in three domains makes the hidden curriculum much broader than the formal or structured component of the curriculum (also called curriculum proper). Instead, it encompasses a whole gamut of learning.

- ***Structured Pedagogies.*** They comprise the most formal elements in the doctoral process, e.g. expected supervisory meetings, required courses for doctoral scholars or their supervisors, research seminars, career or skills workshops and academic conferences. The majority of doctoral scholars aim to capitalise on this institutional provision to support them in their doctoral journeys. Within this category, there remains much flexibility and a range of approaches to how the doctoral provision is conducted, for example. What is worth highlighting is that by intentionally employing further reflection and metacognition, the learning obtained can go beyond what the provision was originally intended for. This is exemplified in Elliot's argument (see Chapter "Developing a Disposition for Harnessing the Hidden Curriculum En Route to Becoming Independent Researchers: The Role of Doctoral Supervisors") where two doctoral scholars' participation in the same workshop could possibly lead to different outcomes. Frick, in Chapter "Creative Supervising/ Supervising for Creativity: Exploring the Hidden Dimensions of Creativity in Doctoral Supervision", also clarified how further reflection on what was learned often leads to what is called 'incubation', which then reinforces knowledge creation.

  As Kobayashi and Berge contend in Chapter "How Humour Can Support Researcher Independence", learning in this sense could mean reflection on interactions between supervisors and doctoral scholars; appreciation derived from sharing humour and laughter encourages the development of trust, security or a sense of belonging. In all these examples, what is apparent is that scholars proactively 'connect the dots' to maximise what their learning experiences can offer them by purposefully thinking about how conventional doctoral learning might bring greater benefits at both personal and professional levels.

- ***Informal Pedagogies.*** Unlike structured pedagogies, there is no pressure to pursue learning in this category. This explains why doctoral scholars individually pursue what they require to satisfy their academic needs, personal or professional aspirations from a wide range of existing opportunities within or outside academia. These learning opportunities might be either apparent or elusive; it is then critical for doctoral scholars to reflect, self-evaluate, search and take the initiative to harness any added doctoral pedagogies that are accessible to them. Guccione in Chapter "Echo-locating a Personalised Route to Independence" discusses how crucial engaging in 'echo-locating' is when navigating one's own circumstances and experiences given the potentially overwhelming learning opportunities scholars are presented with. In this connection, Bryan (see Chapter "Putting the 'Extra' in Extracurricular: Why Going Off-script Is Important For Life After the Doctorate") contends that although extracurricular activities may not be required to meet the criteria for a doctoral degree, extracurricular learning is central to empowering doctoral scholars' overall personal, professional and career development.

  In Chapter "Preparing for the World Outside Academia: Avoiding Organisational Culture Shock", Skakni and Inouye highlight the value of career planning even before the doctoral period is over. Doctoral scholars need to bear in mind ways in which they can simultaneously pursue and nurture career-related learning and activities, particularly since many scholars are now expected to work outside academia. It is important for them to seek a variety of skills, competences and experiences, which will be deemed valuable even by non-academic employers. This is aligned with the reflection that Dangeni et al. offer in Chapter "Peer-mentoring: A Potential Route to Researcher Independence" following three doctoral scholars' stint working together as Graduate Teaching Assistants on the same course. Their experience enabled them to support students registered for this course. What is more, collective working also opened doors in terms of offering and receiving peer mentoring as well as psychological support.

- ***Not Yet Existing Pedagogies.*** Compared to informal resources, pedagogies in this category are mainly pursued, initiated or facilitated commonly by doctoral scholars, supervisors or researcher developers. Needless to say, creating these pedagogies is a prerequisite to accessing them. Compared to the other two pedagogies, the third type requires a lot more reflection as well as investment of time and effort before learning can be harnessed. Creating this type of pedagogy often requires a shared vision and a group decision to co-create the desired learning experience and form both a scholarly and a psychologically supportive community. Very often, there is an extended benefit characterised by the community cultivating and nurturing the group's well-being.

  A number of chapters in this book offer excellent examples for harnessing this category of the hidden curriculum. There are cases of academics forming their respective communities. Makara et al., for example, co-created a Journal Club (see Chapter "'It Is a Nice Way to End the Week': Journal Club as an Authentic and Safe Learning Space"), with a view to engaging a group of doctoral scholars—local and international—to become more reflective and critical readers via participation in informal monthly discussions of selected journal articles. Guerin and Aitchison (see Chapter "Finding Confidence in Writing: Doctoral Writing Groups") also discussed the value of organising doctoral writing groups to support these scholars' collective development but also to build their confidence, work with peers and experience a more holistic approach to researcher development.

  In Chapter "The Interstitial Doctoral Life of #thesisthinkers: When the Hidden Curriculum Might Be All There Is…", Peseta, colleagues and a wider group formed their own platform called #thesisthinkers to foster cross-institutional conversations and support members' collective growth and development. It is worth noting what Peseta asserted, i.e. how creating this new platform is essential, more so because this type of provision does not commonly exist in many higher education institutions. Their group's decision to co-create this group was a response to what was deemed to be lacking in the institutional provision for doctoral scholars and academics.

Across these cases, some characteristics can be strongly observed, i.e. the groups have been transformed into supportive communities cultivating friendship, trust, safe space, constructive criticality, creative thinking, engagement and well-being while supporting scholars' transitions and fostering their researcher identity and independence (Elliot & Makara, 2022). Needless to say, becoming part of a community suggests abundant opportunities for interdependence.

Elucidating Researcher Independence

Whereas the definitions and conceptualisations of the hidden curriculum as described in each chapter have initially been guided by our book's original definition, the various chapters have confirmed, however, what we anticipated from the outset, i.e. that the term researcher independence is a highly contested term. There is a multiplicity of understanding concerning researcher independence since the term can be viewed from multiple angles. This makes this concept elusive, unpredictable and challenging to pin down, which is denoted by the fragmented illustration of 'researcher independence' on the left in Fig. 1.

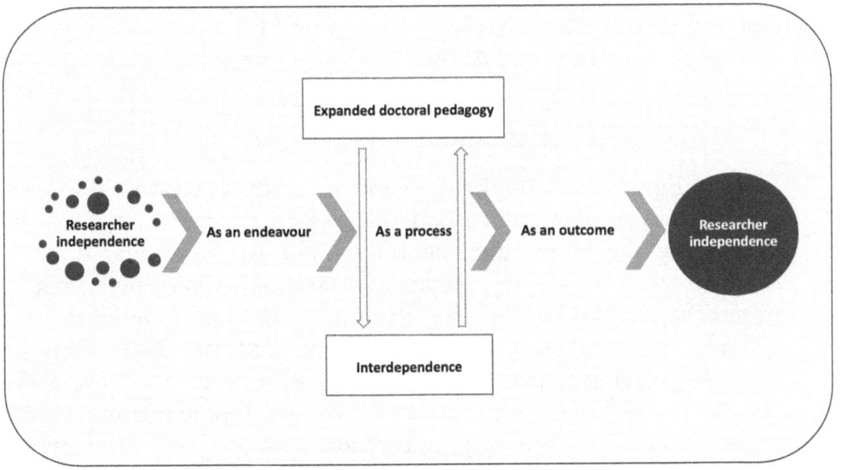

Fig. 1 Three conceptualisations of researcher independence

Gower's chapter is an exploration of the meaning and connotations, general provision and implications associated with understanding researcher independence. In Chapter "The Hidden Meanings of 'Independent Research(er)'", Gower asked crucial questions that are specific to doctoral education, e.g. 'What does independent researcher mean?'—arguing how this term is understood and interpreted is critical. Differences in perception can radically influence doctoral provision and subsequently, the experience of doctoral scholars, and possibly even the examiners' approach to thesis assessment. The complexity entailed in defining researcher independence makes it a more stimulating topic to contemplate—particularly, how it can influence the PhD trajectory by examining how it was expounded in the chapters of this edited collection. It is also worth raising if any of these competing interpretations are more aligned with common wisdom or firsthand observation, backed up by the research literature or empirical studies.

Closer scrutiny of how our various book contributors conceptualised, discussed and exemplified how the hidden curriculum can foster researcher independence led to a greater understanding of the three defining interpretations. Each conceptualisation carries with it a distinct clarity as well as tension emanating from the appreciation and expected execution when promoting independence among researchers. The combined conceptualisations in Fig. 1 clearly suggest that when it comes to fostering and practising researcher independence, the active engagement required from each doctoral scholar starts from the very beginning of the journey, continues throughout and until the end of the PhD finish line, so to speak.

Researcher Independence

- ***As an Endeavour***. In the UK and in many countries, the Vitae Researcher Development Framework (RDF) serves as a guide for describing the knowledge, behaviour and attributes that doctoral scholars need to acquire, pursue and develop to become successful researchers (www.vitae.ac.uk). RDF also commonly underpins the programmes organised by a number of academic institutions for their doctoral scholars. Unsurprisingly, RDF is deemed useful for mapping researchers' reflections of their development using a common language concerning researcher development—arguably an indispensable tool *en route* to researcher independence.

This is demonstrated by Sakurai et al. in Chapter "Mapping the Learning Opportunities of the Hidden Curriculum for International Doctoral Scholars in Japan" based on two international doctoral scholars' case experiences of development via hidden curricular opportunities. Preece, in her capacity as the Researcher Development Programme leader at a British University, introduced a group of doctoral scholars to 'multiple ways of doing' to empower and encourage them to find working preferences that suit their specific contexts. The intention is to develop researcher independence from the outset by urging doctoral scholars to take responsibility for their own research, learning and development (see Chapter "The Dance of Authenticity and Multiple Ways of Doing: Defining a Pedagogy for Accessing the Hidden Curriculum in Doctoral Education and Developing Researcher Independence").

A similar sentiment is expressed by Albertyn in Chapter "Doctoral Intelligence Mechanisms to Illuminate Development Strategies in the Hidden Curriculum". It is important to impress upon doctoral scholars from the outset, what holistic doctoral development requires or what mechanisms can facilitate researcher independence. Doing so can give doctoral scholars a sense of direction from the very beginning in terms of the knowledge and skills that they require to develop.

Not discounting that a PhD is a huge intellectual challenge, it is widely recognised that the doctoral environment generously offers a 'world of opportunities' (Brodin & Avery, 2020, p. 409). It can be further argued that the entire PhD is an invitation to take part in numerous informal interactions subsequently translated into 'mini-learning opportunities' (Elliot, 2022; Elliot et al., 2020, p. 6). As a case in point, exercising independence could mean intentional efforts to become part of a supportive research network where doctoral scholars are given the 'space' to learn, exchange knowledge and grow together as researchers.

Given the non-linear forms of the multiple and diverse opportunities that doctoral scholars are bound to encounter in three, four or more years, they often perceive their doctoral trajectory to be 'very messy' as Aarnikoivu shared in Chapter "Through and Towards An Interdisciplinary Research Community: Navigating Academia as a Lone Doctoral Scholar" (based on her introspection as a doctoral scholar).[1] In general, scholars can expect many serendipitous encounters along the way; their decisions if and to what extent they will take part are arguably key to the quality of their doctoral experience and to their journey to researcher independence. Proactively exploring such opportunities is aligned with McCulloch's (2022) advocacy for making sense of serendipitous occurrences by responding with a 'prepared mind' (p. 258). Doing so can then help transform a serendipitous encounter into a step closer to one's goal of becoming an independent researcher. Bryan contended in Chapter "Putting the 'Extra' in Extracurricular: Why Going Off-script Is Important For Life After the Doctorate" that since independence is much bigger than doctoral scholars' ability to manage a doctoral research project, having a sense of readiness to critically appraise the world around them will help them develop 'intelligent questions and robust methods of answering them'.

Taking the view that researcher independence is an endeavour may also mean using one's agency to plan how doctoral scholars can 'transition to a position of increased researcher agency' by establishing one's own support villages. According to Rainford (see Chapter "Enabling Part Time Doctoral Scholars to Develop Effective Support Villages"), planning to develop support villages needs to take into account future employment, wider disciplines, and writing and publication practices. In this connection, doctoral scholars need not restrict their social support to within the institutional group only. In Chapter "Changing Career Pathways: Making Visible the Employment Destinations and Non-academic Contributions of Doctoral Scholars", Hancock has stressed the importance of tailoring doctoral learning to doctoral scholars' needs and aspirations, with a view to supporting their post-PhD employment and subsequent career pathways. Additionally, Preece maintains that exercising agency is not exclusive to the efforts of doctoral scholars and supervisors. A tailored, authentic and innovative approach to assist doctoral scholars gradually move through stages of researcher development is described and discussed in Chapter "The Dance of Authenticity and Multiple Ways of Doing: Defining a Pedagogy for accessing the Hidden Curriculum in Doctoral Education and developing Researcher Independence".

[1] This is something that may resonate with other doctoral scholars.

- ***As a Process.*** In several chapters, authors stressed the significance of the 'process' component of fostering researcher independence. According to Albertyn (Chapter "Doctoral Intelligence Mechanisms to Illuminate Development Strategies in the Hidden Curriculum"), '[r]esearcher independence is a dynamic process that develops iteratively over time'. Wisker has exemplified this in Chapter "Facilitating Researcher Independence through Supervision as Dialogue" by clarifying the essence of the interactive dialogues between supervisors and supervisees. She stressed that discussions with supervisors have several inherent functions, including academic engagement and cognitive building; feedback giving and pursuit of other creative dialogues—either planned or serendipitous—and often with other colleagues and peers. The dialogue aims to 'demystify both the process and the product', with a view to enabling doctoral scholars to develop independence as well as take ownership of the doctoral journey.

  Along these lines, Frick (see Chapter "Creative Supervising/Supervising for Creativity: Exploring the Hidden Dimensions of Creativity in Doctoral Supervision") drew attention to the supervisors' role in unlocking, stimulating and promoting scholarly independence by using creative supervising and supervising for creativity as pedagogical strategies. These two pedagogies, Frick maintains, are contributory factors towards doctoral scholars' independent thinking to support original knowledge creation. Likewise, Elliot urges doctoral supervisors to impress on doctoral scholars the value of proactively developing a disposition to harness the hidden curriculum using metacognitive strategies (see Chapter "Developing a Disposition for Harnessing the Hidden Curriculum En Route to Becoming Independent Researchers: The Role of Doctoral Supervisors"). Cultivating a metacognitive disposition is argued to be central to becoming competent independent researchers (see Elliot, 2023).

  It is strongly conveyed in Fig. 1's visual representation of the three conceptualisations of researcher independence that 'as a process', development and progress take place within an environment where independence and interdependence are intertwined conceptually and in practice. In Frick's words: 'Knowledge creation is never truly independent, but rather interdependent'.

In addition to recognising that interdependence is the main channel to develop independence, Grant raised a secondary reason for promoting interdependence in Chapter "Midwifing the New: Institutional Leadership for Doctoral Education", which is to address 'a pervasive culture of individualism' in academia. In this respect, Grant has articulated that 'interdependence is not an optional extra to doctoral study but part of the long process of becoming someone who is able to work well with others, to appreciate what others have to offer, and to be able to offer ourselves'.

Makara et al. featured a concrete example of how a small group of doctoral scholars' participation in a Journal Club provided them with an authentic and safe learning space (see Chapter "'It Is a Nice Way to End the Week': Journal Club as an Authentic and Safe Learning Space"). Through regular monthly discussions aimed at developing more reflective scholars and critical readers, the interdependent nature of the Journal Club also promoted socially supportive ways of increasing researcher competence and independence.

- ***As an Outcome***. There are also assertions that researcher independence is viewed as a direct outcome of the doctoral process. Albertyn contended in Chapter "Doctoral Intelligence Mechanisms to Illuminate Development Strategies in the Hidden Curriculum" that researcher independence could be the direct outcome following a holistic development and a good grasp of doctoral intelligence comprising the collective 'knowing', 'doing', thinking' and 'willing' principles. In this framework, Albertyn further suggested that researcher independence is equal to the acquisition of independence in one's thinking.

  Similarly, Guerin and Aitchison demonstrated in Chapter "Finding Confidence in Writing: Doctoral Writing Groups" how doctoral scholars can be assisted to succeed in the research environment. They argued that by activating the hidden curriculum through doctoral writing groups, their acquired insider knowledge helps in developing doctoral scholars' sense of researcher independence. Moreover, as they continue working with peers, they further acquire confidence from observing and experiencing writing success, and this is often translated into other aspects of their doctoral research.

  Given the increasingly complex career trajectories, Skakni and Inouye (see Chapter "Preparing for the World Outside Academia: Avoiding Organisational Culture Shock") asserted the importance of

> doctoral scholars preparing for post-PhD even while undertaking their PhD. This involves building on knowledge and skills to equip them for working either in academia or 'fostering independence in career planning and skill mobilisation', as is increasingly demanded from PhD graduates.
>
> In Chapter "Facilitating Researcher Independence through Supervision as Dialogue", Wisker also cautioned that researcher independence may not be the outcome desired by some doctoral scholars. In this respect, their needs and development during the doctoral journey require consideration. This may explain the different doctoral development trajectories observed among doctoral scholars; for some, independent thinking may 'develop over time, but for some never develop and for others are evident early on'.

Taken together, recognising these three conceptualisations of researcher independence has substantial implications for all key doctoral stakeholders. To begin with, *RI as an endeavour* helps set goals and expectations through the importance placed on becoming an independent researcher. Both researcher independence and doctoral thesis completion are regarded as twin PhD goals, acknowledged by both supervisors and supervisees from the outset. As Gower hinted, this may then form part of the initial discussion and selection process with potential doctoral scholars, particularly in ascertaining their motivations, preferred style of working and aspirations. This also directs all future decisions in relation to what courses, seminars or workshops to engage in. Altogether, RI viewed this way enables doctoral scholars to plan what to prioritise during three or four years of doctoral study.

RI as a process, on the other hand, emphasises two central ideas: (a) acknowledgement of the existence of the expanded doctoral pedagogy via the hidden curriculum (Elliot et al., 2020); and (b) the most efficient road to researcher independence is via interdependent research practices. Doctoral studies' lack of structure warrants greater harnessing of the expanded doctoral pedagogy available through the hidden curriculum to complement and enrich standard doctoral provision. While distinct doctoral pedagogies remain to be discovered along the way, Rainford suggests that it is essential to communicate clearly the value of the hidden curriculum to doctoral scholars. Furthermore, this view of RI not only corrects pervasive preconceptions and possible prejudices surrounding the PhD as research in isolation. It rather illustrates how interactive learning is superior to isolated learning in terms of enriching and widening one's subject expertise, disciplinary knowledge, research approaches, methods and techniques—through cross-fertilisation of ideas via planned and chance interactions (see also Elliot, 2023). It is also worth noting that interdependent learning practices, e.g. being part of a small doctoral community, have an added advantage of tackling a sense of isolation that tends to lead to other issues, and instead helps nurture and preserve its members' well-being.

Finally, *RI as an outcome* has been reiterated in many publications in doctoral studies (Benmore, 2016; Brodin & Avery, 2020; Gardner, 2008; Johnson et al., 2000; Lovitts, 2005; Overall et al., 2011; Savva & Nygaard, 2021; Sverdlik et al., 2018). In this conceptualisation, the idea is that there are qualities that we expect to see in those who have achieved their doctoral qualifications. These are the same qualities that they present or demonstrate to prospective employers as a gauge for meeting job specifications.

Recognition of the increasingly complex post-PhD landscape also prompted many of our book contributors to revisit the parameters for researcher independence. While previously, it was almost assumed that achieving a PhD qualification would lead to working in academia, a combination of factors now suggests that this is no longer the case. In this connection, Skakni and Inouye called for equipping doctoral scholars not only with researcher independence but also with the capacity to exercise independence over career planning and skills transfer.

Of Doctoral Scholars and Redwood Trees

Finally, having explored the dynamics among interlinked doctoral concepts—doctoral pedagogies, the hidden curriculum as a continuum, conceptualisations of researcher independence and interdependence—we would like to leave with you a fitting metaphor.

Redwood trees (or Sequoia sempervirens), a species of coniferous tree, have many characteristics that make them stand out—majestic, stately, sturdy and powerful spring to mind. Various websites featuring redwood trees tell us that they are well known for being beautiful, evergreen and having a 'towering stature'. Redwood trees are claimed to be the largest and tallest trees—capable of reaching 360 ft (or 110 m), which is equivalent to the height of a 35-storey building.

In the right conditions, redwood trees grow fast, approximately two to three feet annually. If you were to walk beneath the canopy of these majestic trees, you could be forgiven for thinking that with such height, their roots are also deeply embedded in the ground in order to remain strong against wind and weather.

By examining redwood trees[2] more closely, one of their most outstanding characteristics is not the depth but the relative shallowness of their root system. Yet, what enables these trees to grow, flourish and stand against storms and even earthquakes is what lies below ground—the very same root system. These roots go down for about 6 to 12 feet before spreading outwards. The redwood's roots comprise tiny roots that not only absorb water and nutrients from the ground but also help the tree to be embedded in the wood itself. The roots stretched out and were intertwined with the roots of other trees. Its distinct root system is considered its strength.

[2] Photo courtesy of Jeff Ma from www.unsplash.com

Carson in her blog post entitled 'What I Learned from a Walk Through the Redwoods' (2018) captured it beautifully:

> These beautiful, majestic giants gain strength from being connected to each other. Their roots intertwine and merge into a connectedness that allows them to nourish each other and hold each other up. When winds and storms, floods and earthquakes come, these trees stand firm and tall because of the connected community their roots have created.

Redwood trees offer an excellent metaphor for all doctoral scholars. In this final reflection, we are aware that growing and thriving within the doctoral forest require a lot more than the standard institutional provision. An expanded doctoral pedagogy is key to maximising institutional provision, but even more importantly, to ensuring that both elements—doctoral research and doctoral development—are given the attention they deserve (Elliot, 2022, 2023). We strongly argue that gradual and sustained growth and development in these two areas serve as a strong foundation for the path to researcher independence. We need to keep in mind though, that just like the redwood trees in the forest, it is critical for both surviving and thriving to be strongly connected and actively participating with the community—harnessing strengths from various stakeholders as well as giving and receiving support in equal measure.

Interdependence is like the intertwined roots of redwood trees that link everything in the doctoral forest together—enriching a meaningful doctoral experience, maximising the chance of a successful and transformative doctoral journey while preserving doctoral well-being through what Kreber (2022, p. 1) refers to as extending 'authentic care' for oneself and others and what Barnacle (2018) refers to as the 'care-full PhD' and doctoral education as developing a capacity to care. We concur with Alves et al. (2023) that persistence to meet the high demands of a PhD does not merely hinge on individual factors; equally, they are very much influenced by meaningful interactions in both academic and social contexts.

As we have demonstrated throughout the book, the hidden curriculum can complement very well the structured doctoral provision, offering its own pedagogies and serving a crucial role in developing competence and independence among doctoral scholars. To achieve such a level of researcher independence, interdependence seems to be a catalyst. We would like to argue that interdependence goes beyond that, i.e. it also assists in nurturing, cultivating and safeguarding doctoral well-being.

References

Alves, P., Lopes, A., Correia, R., & Menezes, I. (2023, June 6). The Interrupted Journey: Factors and processes related to withdrawal, re-enrolment and drop-out from doctoral education. EARLI SIG (Researcher Education and Careers) 24 Webinar.

Barnacle, R. (2018). Research education and care: The care-full PhD. In S. Bengtsen & R. Barnett (Eds.), *The thinking university. A philosophical examination of thought and higher education* (pp. 77–86). Springer.

Bengtsen, S. (2016). An exploration of darkness within doctoral education. Creative learning approaches of doctoral students. In C. Zhou (Ed.), *Handbook of research on creative problem-solving skill development in higher education* (pp. 260–282). IGI Global.

Bengtsen, S., & Barnett, R. (2017). Confronting the dark side of higher education. *Journal of Philosophy of Education, 51*, 114–131.

Benmore, A. (2016). Boundary management in doctoral supervision: How supervisors negotiate roles and role transitions throughout the supervisory journey. *Studies in Higher Education, 41*(7), 1251–1264. https://doi.org/10.1080/03075079.2014.967203

Brodin, E. M., & Avery, H. (2020). Cross-disciplinary collaboration and scholarly independence in multidisciplinary learning environments at doctoral level and beyond. *Minerva, 58*, 409–433. https://doi.org/10.1007/s11024-020-09397-3

Carson, M. (2018, July 16). What I learned from a walk through the redwoods. Their story is our story. https://tsosrefugees.org/blog/2018/07/16/what-i-learned-from-a-walk-through-the-redwoods

Elliot, D. L. (2022). A 'doctoral compass': Strategic reflection, self-assessment and recalibration for navigating the 'twin' doctoral journey. *Studies in Higher Education, 47*(8), 1652–1665. https://doi.org/10.1080/03075079.2021.1946033

Elliot, D. L. (2023). *Navigating your international doctoral experience (and beyond)*. Routledge. ISBN 9781032220505

Elliot, D. L., Baumfield, V., Reid, K., & Makara, K. A. (2016). Hidden treasures: Successful international doctoral students who found and harnessed the hidden curriculum. *Oxford Review of Education, 42*(6), 733–748. https://doi.org/10.1080/03054985.2016.1229664

Elliot, D. L., Bengtsen, S. S. E., Guccione, K., & Kobayashi, S. (2020). *The hidden curriculum in doctoral education*. Palgrave Pivot. https://doi.org/10.1007/978-3-030-41497-9

Elliot, D. L., & Makara, K. A. (2022). An online community of international scholars: Enabling spaces for reciprocal academic and psychological support. *Innovations in Education and Teaching International, 58*(6), 693–703. https://doi.org/10.1080/14703297.2021.1991424

Gardner, S. K. (2008). 'What's too much and what's too little?': The process of becoming an independent researcher in doctoral education. *The Journal of Higher Education, 79*(3), 326–350.

Johnson, L., Lee, A., & Green, B. (2000). The PhD and the autonomous self: Gender, rationality and postgraduate pedagogy. *Studies in Higher Education, 25*(2), 135–147. https://doi.org/10.1080/713696141

Kreber, C. (2022). On a pedagogy of authentic care in postgraduate research supervision. *Studies in Higher Education*. https://doi.org/10.1080/03075079.2023.2215805

Lovitts, B. E. (2005). Being a good course-taker is not enough: A theoretical perspective on the transition to independent research. *Studies in Higher Education, 30*(2), 137–154. https://doi.org/10.1080/03075070500043093

McCulloch, A. (2022). Serendipity in doctoral education: The importance of chance and the prepared mind in the PhD. *Journal of Further and Higher Education, 46*(2), 258–271. https://doi.org/10.1080/0309877X.2021.1905157

Overall, N. C., Deane, K. L., & Peterson, E. R. (2011). Promoting doctoral students' research self-efficacy: Combining academic guidance with autonomy support. *Higher Education Research & Development, 30*(6), 791–805. https://doi.org/10.1080/07294360.2010.535508

Savva, M., & Nygaard, L. P. (2021). In M. Savva & L. P. Nygaard (Eds.), *Becoming a scholar. Cross-cultural reflections on identity and agency in an education doctorate*. UCL Press.

Sverdlik, A., Hall, N. C., McAlpine, L., & Hubbard, K. (2018). The PhD experience: A review of the factors influencing doctoral students' completion, achievement, and well-being. *International Journal of Doctoral Studies, 13*, 27.

Vitae. (n.d.). About the vitae researcher development framework. Vitae. https://www.vitae.ac.uk/researchers-professional-development/about-the-vitae-researcher-development-framework

Wisker, G., Robinson, G., Trafford, V., Warnes, M., & Creighton, E. (2003). From supervisory dialogues to successful PhDs: Strategies supporting and enabling the learning conversations of staff and students at postgraduate level. *Teaching in Higher Education, 8*(3), 383–397. https://doi.org/10.1080/13562510309400

Index

D. L. Elliot et al. (eds.), *Developing Researcher Independence Through the Hidden Curriculum*,
https://doi.org/10.1007/978-3-031-42875-3

SPRINGER NATURE

GPSR Compliance

The European Union's (EU) General Product Safety Regulation (GPSR) is a set of rules that requires consumer products to be safe and our obligations to ensure this.

If you have any concerns about our products, you can contact us on ProductSafety@springernature.com

In case Publisher is established outside the EU, the EU authorized representative is:

Springer Nature Customer Service Center GmbH
Europaplatz 3
69115 Heidelberg, Germany